Smoke Ash and Steam

Steam Engines Used on the West Coast
How They Worked

R Shuet

Canadian Cataloguing in Publication Data

Sheret, R. E. (Robin Edward), 1932-
Smoke ash and steam

Includes bibliographical references and index.
ISBN 0-921107-04-8

1. Marine engines--Pacific Coast (B.C.)--History. 2. Marine engines--Pacific Coast (U.S.)--History. I Title.
VM731.S53 1997 623.8'722'0979 C96-910863-X

First Edition 1997

Western Isles Cruise & Dive Co. Ltd.
2962 Leigh Rd., Victoria B.C. Canada V9B 4G3

Printed in Canada

Smoke Ash and Steam

Steam Engines on the West Coast of North America

Robin E. Sheret

Dedicated to all the steamboat men that gave a green kid a start in life. To all the men and women that work so hard to preserve the remaining steam boats and engines.

WESTERN ISLES

Western Isles Cruise & Dive Co. Ltd

Acknowledgments

Many thanks to all the people that assisted us with this project. Without the museums, societies and marine parks mentioned below we could not have put this book together.

San Francisco Maritime National Historical Park (National Park Service}
Oregon Maritime Center and Museum
Virginia V foundation
UASBC (Underwater Archeological Society of B.C.)
Anacortes Museum
SS Moyie National Historical Site
Samson V Maritime Museum
Whatcom Maritime Historical Society
National Liberty Ship Memorial
SS Master Society
SS Sicamous Restoration Society
Museum Fort St. John
Wells Books (Victoria)
Saanich Historical Artifacts Society

Chief Editor - Joan Sheret

The following individuals have assisted us with their input:
Ted Miles, R.J. Smith, Diane Wells, Fred Anset-Beale.

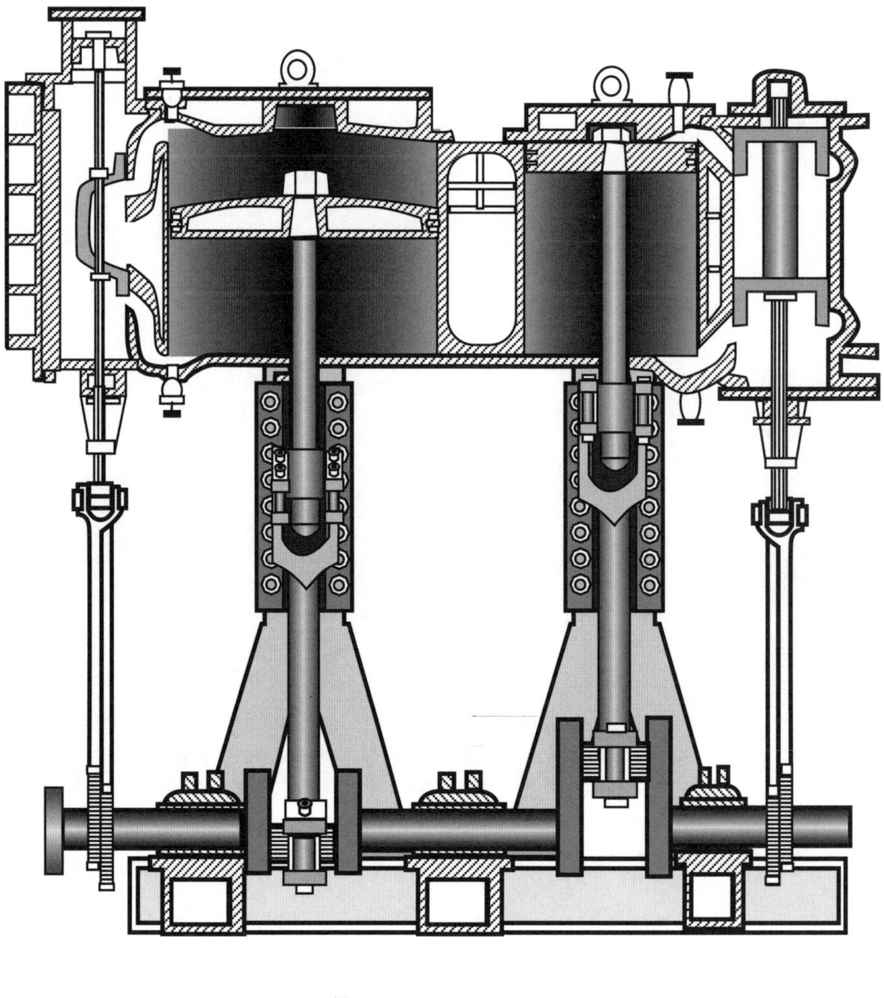

Contents

Introduction

We have spent the last few years collecting information on steam engines used on the waterways of the West Coast of North America. As someone who started his career on steam vessels I found this very interesting and a great learning experience. We looked at existing equipment, studied many old manuals dating back to the middle of the 19th century and talked to engineers and archeological divers.

This book is meant for the person interested in coastal shipping. Our apologies to engineers who may find it a bit simplistic. You may find we have repeated some things, such as the jet condenser, but we imagine most readers will look at an individual chapter rather than read the book from cover to cover.

There have been many books written on the big liners but it was the smaller coastal ships that opened up the West Coast. Despite this it was the development of the steam engine on the East Coast and in Europe that affected our shipping. For this reason we have discussed these developments in some detail.

The date that a particular piece of equipment came into use is often hard to establish. At first there were all sorts of things tried. Some of these worked well in a particular area and not in another. Sometimes it took years of trial and error before a concept was proven reliable for general use. The screw (propeller) driven ship was an example of this. In addition, records were often not too well kept.

Terminology differed in time and area, sometimes in the same engineering manual. We have tried to use the most common terms used in the U.S. and Canada. Anyone reading the specifications of a particular vessel should be careful of this. There have been a number of names used for the same type of engine. Horsepower can lead anyone astray especially as nominal horsepower is often quoted.

We hope you find this book as interesting as we did doing the research. This is not the end of our curiosity into coastal subjects. We hope to keep on discovering information about the tugs, ships and seamen that developed the Coast. I am proud to have been a part of the marine industry it kept me alive for years.

How to use this book

Each chapter is intended to stand on its own. For this reason you may find some information repeated. If more information on a particular subject is required, such as condensers look in other chapters. If the reader is not familiar with steam we suggest reading chapter 2 first.

Where did we get the information?

We have been collecting information for this book for a number of years. It did not start out as book material. In fact we were working on another book when this started. I started dragging out the old engineering books after diving on a wreck of a steamer. We had some problems trying to identify parts. We soon found other people were interested in the subject. The more we traveled and studied the more interesting it became. It soon became apparent that each individual engine, with the exception of some built during the war, was different. We relied heavily on old engineering manuals, dating back to 1855, for information and the opinions of the time. Unfortunately space limits the amount of information we can put in a book.

The information was collected from:

1 - Working on steam boats
2 - Talking to wreck divers (UASBC)
3 - Diving on wrecks
4 - Talking to engineers and others knowledgable in steam history
5 - Looking at existing steam engines from Ontario to B.C. and down the coast to San Francisco.
6 - Engineering manuals dating back to 1855
7 - Books on nautical archeology
8 - Photographs of steamers
9 - Books on nautical history

Western Isles Cruise and Dive Co. Ltd.

Victoria B.C.

The Steamboats

Chapter 1

Steam Ships On The West Coast

The Canadian Department Of Transport Lighthouse Tender Estivan

The Estivan, originally a coal burner, was the main lighthouse tender for the Department of Transport (now Coast Guard) on the B.C. coast for years. She worked from Egg Island south, repairing and replacing buoys. She supplied the lighthouses and repaired navigation markers. This was before the days of helicopters. All the material supplies went ashore by boat. At first the boats were rowed into the lights before motor launches were used. She would load bunker coal and coal in Union Bay or Nanaimo..
1161 Gross Tons, Registered length 212 ft. Power 174 nhp
While this ship was a government vessel she was similar to many coastal freighters used all over the coast.

Steam Boats on the Pacific Coast

When we look at the history of steamboats and their engines on the west coast of North America from 1830 to 1955 we also must look at what had gone before in Europe, East Coast, Great Lakes and the Mississippi Ohio river system. The Mississippi Ohio Rivers and all the tributaries were known as the western rivers but to avoid confusion we will call it the Mississippi River. These areas all had an influence on the vessels operating on the coast, the rivers and lakes especially in the early days. Local vessel owners soon adapted this experience to the unique conditions found in the west. When steam technology improved it was not always embraced wholeheartedly. For this reason sidewheelers and other older types of vessels were still built long after compound and triple expansion engines had become the norm elsewhere.

Space permits us to mention just a few of the highlights of coastal history. We have only scraped the surface of this interesting subject.

The number of steamers steadily increased to service the developing communities on the coast after the arrival of the famous Hudson Bay Company ship Beaver. Steam was economically viable in coastal areas in Europe and North America long before it replaced sail on the open ocean routes of the world. The sailing ship whether square rigger or schooner, remained the most economical and reliable means of transportation on long hauls till the '90's. Some of the early deep water steamers were converted to sail as they were not economically viable. Sail and steam worked together for years. Side paddlewheels were used on a lot of early passenger and freight ships. Paddlewheelers stayed in use long after propeller-driven vessels were developed because they were faster and more reliable. The engines and propellers on the first screw-driven vessels were of questionable quality. For years ocean-going steamers had sails just in case the engines quit or they ran out of bunkers.

A number of vessels in the early days met an inglorious ending. Competition was very stiff, regulations were lax or ignored, overloading was normal and aids to navigation were scarce or nonexistent. The U.S. did not have many laws governing steamboats till 1852. The West Coast with its many islands, rocks, sudden storms and strong currents is a perilous area for navigators. Many ships wound up on the rocks, often with great loss of life. It's to the credit of the crews there were not more disasters.

The Beaver, the first steam ship on the West Coast of North America, equipped with side-lever engines arrived on the Columbia River in March, 1836 just 29 years after Fulton's Clermont became the first commercial steamboat and 20 years after the Washington, the first successful Mississippi river boat. In 1848 the first steamers arrived in California from the East Coast. In 1850 the first sidewheeler was built on the Columbia River.

By 1853 the Otter, the first propeller-driven steamer arrived on the coast. The Otter operated from Alaska to San Francisco. In the same year, the sidewheeler Fairy was brought to Puget Sound for the mail run between Seattle and Olympia.

The history of steam on the West Coast could be divided into three basic areas. They are, Southeastern Alaska, British Columbia and Puget Sound, Juan de Fuca to Panama, rivers, lakes and harbors.

Pacific North West - British Columbia and Puget Sound

The area from Puget Sound to Alaska has strong tidal currents, variable winds, calms, sudden gales and many rocks and reefs. Sailing vessels were not too practicable for inshore operations under these conditions and often had to be towed. Many a sailing ship perished on the West Coast of Vancouver Island and in Juan de Fuca Straits. The settlements on inside passage of B.C.and Puget Sound were originally developed with the aid of steam vessels. Passenger vessels, tugs and coastal freighters all played a part in this.

The H.B.C. vessel Beaver (1836), operated in all these capacities. The H.B.C. vessel Otter (1852) the first propeller-driven ship was used for trading, freight and passengers.

Puget Sound started getting vessels from the south to service the early settlers and the lumber industry. Columbia River interests had control over parts of the shipping in the Sound for years. Despite this many independently owned vessels serviced the Sound. These ships both large and small became known as the mosquito fleet. Early vessels like the Fairy, used on Puget Sound runs had problems. Her boiler blew up in 1857 sinking her. The propeller-driven Major Tompkins arrived in the Sound in 1854 to carry mail, passengers, and freight between Olympia, Seattle, and Victoria. She was 93ft. (28.4m) long of 150 net tons, built in 1847. She was not noted for her speed and ended her days stranded off Victoria in 1855.

Sidewheelers, sternwheelers and screw-driven vessels were all used in the Sound and inside passage for years. Some of the paddle boats were in use long after screw-driven vessels had taken over all the deep water and open water coastal trades.

The H.B.C sidewheeler, Labouchere arrived in 1859 to start trading in the north. She was also used on the southern runs.

In 1858 when gold was discovered in the interior of what is now British Columbia ships started running between Victoria and New Westminster. By this time the U.S. had developed a fleet of steamers and some of the first ships on this run were American operating under licence. The Beaver and the Otter under the Hudson Bay flag were put on this run. Victoria started to build ships for this run. The Hudson Bay purchased the Enterprise so the Otter could be put back on the northern run. The Enterprise was built in San Francisco and was equipped with a walking beam engine. The marine operations of the Hudson Bay Company were later to become the Canadian Pacific Navigation Co. Their vessels were built in Victoria, U.S. and England. The company operated passenger and freight services on the B.C. coast until the C.P.R. took it over in 1901. A number of C.P.R. vessels were built in the U.K. as were a number of ships owned by other companies.

Other coastal freight and passenger ships serviced the many small communities developing along the coast. Canneries were built in the inlets in B.C. and S.E. Alaska. Supplies and passengers were delivered by passenger and freight boats on the way up. Fish, canned fish, fishmeal and passengers were picked up on the way south. The crews worked the freight in each port whether it consisted of one package, a cow, or tons of supplies or products. Many of the coastal passenger vessels were miniature ocean liners with elaborate passenger accommodations. Most of them also had a hold for freight. This type of British style passenger freighter was more common in B.C. than in the United States.

On the B.C. coast a number of companies operated coastal steamship operations. The C.P.N. later C.P.R. operated passenger and freight vessels to the West Coast of Vancouver Island, New Westminster, Seattle and northern B.C. ports. Many of their ships were built in the UK. They also liked to use British officers.

The Grand Trunk Pacific Co. started running a similar type of ship from 1910 to 1975. Most of their ships and some of their crews came from the U.K. They operated from Vancouver to Prince

Rupert as well as Seattle and Alaska. The Grand Trunk Railway started the city of Prince Rupert in 1908 as their western terminus.

From 1890 to 1959 Union Steamships connected Vancouver with every inlet and town on the upper coast. There were other competitive companies such as Waterhouse and Northland operating many ships on the coast.

The ferry services in Puget Sound and in B.C. eventually took over much of the marine passenger trade. The Seattle designed for wagons and passengers was the first ferry on the Sound in 1889. Black Ball started operating in the 1930's. After 1939 a number of the San Francisco ferries wound up in the Sound. The present B.C. Ferry system started in the 1960's but they were all diesel-powered ships.

Tugs on both sides of the border moved logs to the new mills that were springing up in the area and transported lumber to markets in the larger cites. Towing sailing ships into ports from Juan de Fuca and across the Columbia River Bar became a

The steam tug Spray built in 1907. Owned by Victoria Tug. Compound engine 221/2 nhp. Reg. length 81 ft. 118 Gt. Scow towing tug (mostly coal)
Photo by R. Richardson

very competitive business. From this a type of North American coastal tug developed that was quite unlike European designs. The crews also became specialists. You might be an elite master of deep water ships but totally useless on a tug. Each area developed its own towing skills. The operation of the stern wheel tugs on the Columbia and many lakes and rivers further north, was completely different from the operation of the B.C. and Puget Sound boats.

The early steam tugs were very low powered by today's standards and manoeuvering them and a tow required a lot of skill. The first ones often operated as passenger and freight ships in addition to towing. Small tugs often towed more than one sailing ship at a time. This sometimes led to disaster if the wind came up fast. There were no towing winches on the the early tugs making towline handling a problem. On photographs of some of the old ones you will see tow bits on top of the housework.

CPR passenger-freight stern wheeler Okanagan at Kelowna 1919. 1077.78 GT. 2 horizontal engines 96"x22", 560 IHP Built 1907.
Photo by Parlby

This must have created no end of troubles. There were some paddle wheel tugs at first but most of them were propeller driven.

When more mills were built on the coast more tugs were built. In addition tug and barge operations did the same job as coastal freighters did in other parts of the world.

As the tow boat industry developed so did the tugs. Some of the early tug boat companies were Preston Manning, Foss, Cliff, and Puget Sound Towing. They became more powerful and equipment such as steam-driven towing winches were installed.

Juan de Fuca to California

Below the mouth of Juan de Fuca strait there was not much protection from the Pacific gales. Until a vessel reached San Francisco most of the harbors had to be entered after crossing a bar such as Grays Harbor or the Columbia River. Anyone that has crossed the Columbia River Bar in bad weather will know what we mean. This could be

Model of a Columbia River sternwheeler. Portland Maritime Museum. Typical of passenger-freight vessels operating on the lower Columbia River

disastrous with much of a sea running. The ships operating along this part of the coast were usually larger than the ones working the Sound or the inside passage.

When California became part of the U.S. in 1848, steam ships were employed under government subsidy, taking passengers and mail from the Isthmus of Panama to San Fransico and Astoria. It was a long trip across the U.S. by land but by taking passengers down the East Coast by ship to Panama and up the West Coast by ship the time could be reduced. This became especially useful after gold was discovered in California.

Three sidewheelers equipped with side-lever engines were built in New York for the Columbia River to Panama run. The first to arrive was the 199 ft. (60.6m) California in 1848. She was followed by the Panama and the Oregon. As the arrival of steamships preceded the construction of machine shops and larger foundries, they had to bring spare parts with them.

The lack of shops and tradesmen took a few years to rectify. It was not too long till shipyards and engineering works were established. They were soon able to build river and smaller coastal vessels including the machinery. This west coast shipbuilding industry developed in most major coastal centers.

Many vessels ran up from San Francisco to the Columbia River, Puget Sound, Victoria, and Alaskan ports. At first steam concentrated on the passenger trade. Sail was more than competitive on freight but steam slowly took over all coastal operations. The first ones were wooden paddle wheelers but they were eventually replaced by wood and iron propeller-driven ships.

Steam schooners were used for moving lumber and other cargos to San Francisco and other California ports. At first they were big wooden sailing schooners with engines added. After the 1880's they were designed as steam ships and lost their

sailing ship lines but remained constructed of wood. They were built in Northern California and they could be found working anywhere on the coast. In the 1920's steel steamships built for WW I replaced them.

Alaska

The vessels supplying Alaska have a lot to contend with. The ship can encounter poor visibility, sudden gales, ice, snow and heavy seas.

The first regular run to Alaska from U.S. ports was started in 1867. The ships sailed from Portland, Oregon at first. Soon ships were sailing from San Francisco as well.

In 1875 the Oregon Steamship Co. took over the mail contract. In 1881 the Pacific Coast Steamship Co. put three vessels on the run to com-

pete with Oregon Steam with stops in Puget Sound and Victoria. In 1894 Alaska Steam was incorporated in Port Townsend.

The 1897 gold rush in the Yukon brought relief from a depression and a revival of the shipping industry on the West Coast. Many ships were brought back into service for runs to Alaska. Some of them were old and rotten and would not be allowed to operate under other circumstances. The ships sailed from Seattle, San Francisco and Victoria loaded with supplies and miners.

After the gold rush the coastal towns were served by Alaska Steam along with a few other companies. Summer cruises were becoming fashionable.

Canadian ships became involved with the Alaska trade too. The Islander made some trips before the gold rush. As soon as the gold rush started she along with other ships made steady runs to Alaska. After the rush the C.P.R., C.N.R., and Union Steamships all ran ships to Alaska in the summer.

Alaska Steam eventually became a freight only operation. After WWII old Liberty ships were put on the run but they could not get into some of the ports. Sealand used faster ships. The earthquake in 1964 and changing demands ended Alaska steam in 1971. Tug and barge operations

The last of the Canadian coastal passenger ships

For years the coast was served by passenger steamers that resembled small ocean liners. The last of British Columbia's steam passenger ships was the Princess Marguerite.She was built in Scotland 1949 for the CPR. She was powered by turbo-electric engines developing 7750 shp on each engine. Speed on trials 23 knots. Length 373.7 ft. overall. 5911 GT. She ended her career on the Victoria to Seattle run under the B.C. Steamships flag. Her last run was in September 1989. This ended the coastal steam era.

The Prince George was a Canadian National Ltd. coastal passenger ship. Main run was from Vancouver to Prince Rupert and Skagway Alaska. Length approximately 335 ft. Built in Canada.

The Yukon Star. She was originally a castle class corvette. Built in the U.K. 1943. Union Steamships converted her to a passenger -freight vessel Camosun III in 1946. 1835 GT. Length 235.7 ft. Triple expansion steam engines producing 2800 IHP.

The Island Commander as shown here was a diesel tug but was originally a steam ship. Many of the West Coast vessels had a number of occupations. She was built in the U.K. as a steam trawler. She had an iron hull.

took over most of the heavy freight and rail business. Planes and ferries took over the non-cruise passenger business.

Rivers - Lakes and Harbors

While deep sea ships and coastal vessels can go part way up the major rivers most river boats have to be shallow draft and manoeuverable. The crews must be a special breed as well. They needed to know the ever changing river, load heavy cargos and bunkers without machinery and be able to turn their hand to any job that came up.

The settlements along the rivers and lakes in the interior of B.C. depended on water transportation up to the 1950's. Some areas such as the lower Fraser still depend on it. The Columbia River always used water transportation. The Sacramento River was served by boats for years.

Sternwheelers operated all over the coast and on rivers and lakes for a many years. While the original boats were copied from the Mississippi, the hull design on the sternwheelers used in B.C. and on the Columbia River were usually quite different from the vessels used on the Mississippi. Most western sternwheelers had a single stack rather than the twin stacks. The sternwheelers on the West Coast had finer lines forward than did a lot of Mississippi boats. Unlike the Mississippi boats, sidewheelers did not work out on the fast running West Coast rivers. Larger sternwheelers were used on a lot of the lakes for passenger and freight. The rivers and lakes generally used sternwheelers for tugs though some conventional tugs were built for the lakes. Until the 1850's San Francisco had some sidewheel harbor tugs.

The first steam boat in San Francisco was the Russian built sidewheeler Sitka from Alaska. She arrived by sailing ship in 1847. She was 37 ft. (11.3m) long and not too powerful. When gold was discovered in 1848 the riverboats soon followed. They worked the Sacramento and other rivers. At first they were shipped in pieces from the east but local enterprises soon took over the construction. In the beginning they were a combination of Hudson River and Mississippi designs. A lot were sidewheelers with low pressure engines. By the turn of the century most were sternwheelers with horizontal high pressure engines and a single stack. At first the sternwheelers used high pressure engines similar to the Mississippi river boats. They used the traditional poppet valves but by the 1880's the slide valves were in general use.

There were a rash of boiler explosions from 1853 until the 1890's. A lot of this was blamed on boat races and carelessness but some of it was probably due to the boilers and other equipment the engineers had to work with.

In addition to the fancy passenger vessels there were more utilitarian freight boats built for river work. Some towed small barges from the king post or tall towposts after the mast. These vessels operated in the area until the 1930's until roads and railroads put them out of business.

San Francisco developed a fleet of ferries to move people across the bay. The first regularly scheduled ferry started in 1850. They were paddle wheelers but a lot of the later ones were screw driven. When the bridges were built in the 1930's the number of ferries was reduced. Some did operate until the '50's moving rail cars.

The Columbia River was developing its own fleet of tugs and passenger vessels just as on the B.C. coast and in the Sound. In 1846, Oregon became part of the U.S. The first American steamer in the Columbia was the propeller-driven Massachusetts in 1849. The first sidewheeler built was the 90ft. (27.4m) Columbia in 1850. River boats at first were equipped with eastern engines and boilers. The Mississippi style of boat or propeller-driven boats did not work out too well on the

A model of a steam schooner in the Maritime Museum, National Historical Park, San Francisco. The housework has been cut away to show the engine room. Vessels such as this operated from California to Puget Sound.

The side-wheel ferryboat Eureka built in 1890. Walking beam steam engine. Length 277 ft. GT 2420, 1500IHP. San Francisco MaritimeNational Historic Park

S.S. Jeremiah O'Brien.
Ship building was a major industry on the West Coast during the war. The liberty Ship as she was in WW II.
Length 441.5 ft. About 7500 GT. 14000 displacement tons. They carried around 9146 tons of cargo.
Triple expansion reciprocating engine 2500 IHP.
Alaska Steamships used Liberty Ships after the war.
The National Liberty Ship Memorial, San Francisco

The R.M.S. Niagara departing Vancouver B.C. At one time there were many liners operating out of West Coast ports. She looks like a big version of the British built coast steamships. Photographer and date unknown.

The Mills Trader.
This was a type of WW II freighter that sailed under several flags after the war. They had reciprocating engines and were oil burners.
Originally a Park Ship built in Vancouver.

river. The Jennie Clark (1855) was built as a sternwheeler enabling it to operate under the difficult conditions the river presented.

The first sternwheelers on the Fraser River were American working under licence so the development of these vessels followed the U.S. model rather than the European. After gold was discovered in 1857, American sidewheelers and sternwheelers appeared on the Lower Fraser. These vessels would run between Victoria, New Westminster and Yale. Like the San Francisco boats, they had the same problem with boiler explosions. Victoria soon started building boats for this trade.

The crown colony of British Columbia was formed in 1857. The population increased quickly in the lower mainland.

By 1865 the gold had petered out and the amount of river boats on the lower Fraser diminished. In 1866 Vancouver Island and British Columbia amalgamated and became part of the Dominion of Canada in 1871. During this time vessels built in the U.S. were brought to B.C. and sometimes went back to the U.S. again.

On the upper Fraser sternwheelers operated from 1863 to 1921. The boats would be built in the area but the machinery would be brought up by wagon. By this time the hazards of navigation were far greater than the chance of a boiler explosion.

Up north the Skeena River had the first sternwheeler arrive from Victoria in 1864. As more power was needed a new boat was built in Victoria in 1866. From then on most rivers and lakes had sternwheelers. When the Grand Trunk Pacific, later C.N. was building their rail road along the Skeena, in 1908 they built three sternwheel freight boats to move supplies up river.

The horizontal type engines were often salvaged from older ships. The boats usually had finer lines than the typical Mississippi and Ohio boats. The boiler was forward and the engines aft to balance the boat.

There were one or two boats on the Yukon river before the gold rush but after gold was discovered boats were imported or built as fast as possible. The Yukon River communities used boats for many years after the rush.

General

From the earliest days of steam on the coast the vessels were owned and operated by many highly competitive small operators and big companies like the Hudson Bay Co. and the Pacific mail line. The big companies eventually took over most of the passenger trade. Some of the vessels were old castoffs from other parts of the world while others used the best technology available. During this period there were many types of engines and vessel design. If a ship was wrecked or damaged the machinery was often salvaged and used in another vessel. A ship or tug might have sailed under U.S. and Canadian (British) flags in its career. For this reason when looking at the history of a vessel the machinery may have originated in Europe, the U.S. or Canada. As shipyards and machine shops devel-

Virginia V - Seattle
The last of the Mosquito Fleet. She is typical of the small passenger freighters that operated on Puget Sound and in B.C. They were the main form of transportation in small communities for years.
Built 1922
Engine 1904 from Virginia IV triple expansion - 400 IHP.
Virginia V Foundation.

Tug Hercules, Typical of a older style of tug. A high wheelhouse for good visibility when handling barges alongside. The high wheelhouse was added on after it started working for the Western Pacific Railroad.
Built in 1907. Triple expansion 500 IHP. Length 151 ft. GT 409
San Francisco Maritime Historical Park.

oped, the design of a lot of the ships and their machinery evolved into a west coast style.

After the 1870's engines and ship building became more reliable. Most of the coastal vessels were fitted with reciprocating engines. Paddle wheel boats operated on Puget Sound for some time after the propeller ships had taken over other trades. On the rivers and lakes, the sternwheelers and the HP horizontal engine were still used after steam had nearly disappeared on the coast. Due to weather conditions and the strong tidal streams on the coast sidewheelers disappeared long before they did in Europe. On some of the coastal passenger vessels turbines replaced the "up and downer". Eventually the more efficient diesel slowly replaced steam on the coast completely. Due to different regulations steam lasted longer on tugs and smaller ships in Canada than in the United States.

As time went on shipbuilding became a major part of the West Coast economy. This ship building experience was well used in WWII. All types were built from wooden tugs and freighters to Liberty Ships, Victory Ships, and Park Ships.

Passenger vessels, freighters and tugs were not the only ones to use steam. Whalers, fishing vessels and government ships were often steam powered. While the later whalers used in B.C. were the traditional hunter vessels, in California steam was used on sailing ships around the turn of the century. Some of these vessels originated in Europe. If it had not been for the steam ships and the steamboat men development of the west coast of North America would have been a lot slower.

Steam
What Made It Work?
Chapter 2

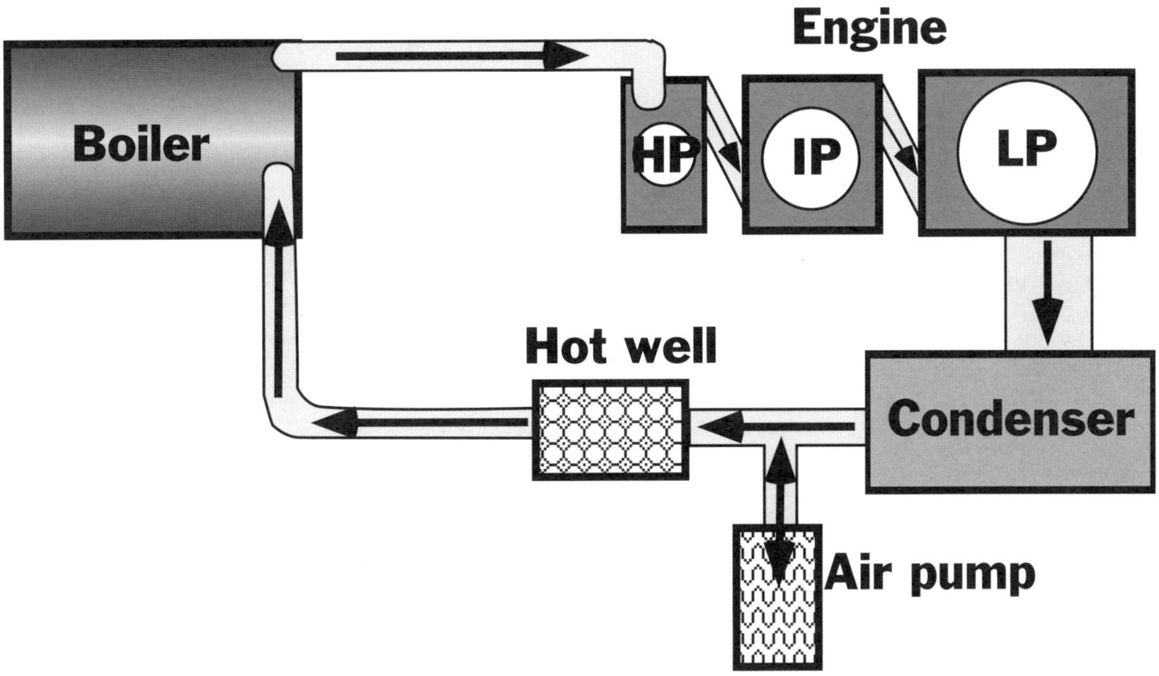

On most steam engines large or small the steam is generated in the boiler. From the boiler it goes directly to the engine. In later engine rooms a separator to remove any water might be installed. From the engine it goes to the condenser. The vacuum necessary for most engines to operate efficiently is created in the condenser. Some high pressure engines such as river boat engines did not have a condenser. The air pump removes the air and water from the condenser and pumps the warm water to the hot well. The water from the hot well (feed water) is injected back into the boiler.

Part one is a thumbnail history of steam.

Part two shows things that affect most steam engines.
The reader may want to refer back to this section to better understand descriptions of individual engines.

Part 1 - Steam Engine

There have been many arguments as to who actually invented the steam boat. The first practical steam powered vessel in the U.S. was built by Robert Fulton in 1807. The date of a particular engine or technology is often hard to establish. A concept may have been around for years before it came into general use. It may have been accepted over time in one area and not another.

The steam engines used on the West Coast were originally developed in Europe, East Coast, Great Lakes and the Mississippi (western rivers). Some, but by no means all, of the early problems with steam had been overcome by the time steam arrived on the West Coast in 1836. Changes in ships and engines in Europe and on the East Coast affected the West Coast in various ways and times.

One of the earliest practical engines, after the original Boulton and Watt engine was the crosshead type engine. This in turn was replaced by the side-lever engine and the vertical beam or "walking beam" engine. The British preferred the side-lever engine with its lower center of gravity. The side-lever engine was used on both sides of the Atlantic for ocean-going ships. The American vertical beam engine was used in protected waters for years. However it was used on some ocean-going vessels for a while. The steeple engine, another low pressure type, was used on some early steam ships. The U.S. Navy used a low pressure type of inclined engine on some of their ships. Another engine invented in England and used in North America in the 1840's, was the oscillating engine. These were basically all low pressure non-compounding paddlewheel engines. Boilers of the day could not handle much pressure. These engines were terrible fuel hogs. They needed a large engine room crew (the black gang) to feed the many boilers and handle all the fuel. At the same time these engines were being used, a type of high pressure engine was being developed on the Mississippi and Ohio River systems.

Before higher pressures could be achieved, the technology used in engine making had to improve as well. In the 1850's hemp was used for packing pistons instead of using piston rings.

While the engines used for propulsion were designed for marine use a lot of the auxiliary equipment such as winches were developed for stationary engines ashore. Hoisting engines were adapted for a variety of uses including early towing winches. Along with the development of the early steam engines, deck winches, steering engines, cargo winches, anchor winches and steam driven capstans evolved. A dive on a wreck of a steamer built in the 19th century would not show much difference from equipment built in the 20th century. One of the advantages of steam is that there is power to any place you can run a steam pipe.

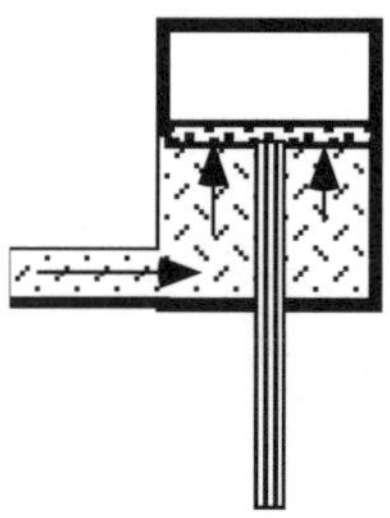

Single Action Steam used on only one side of piston

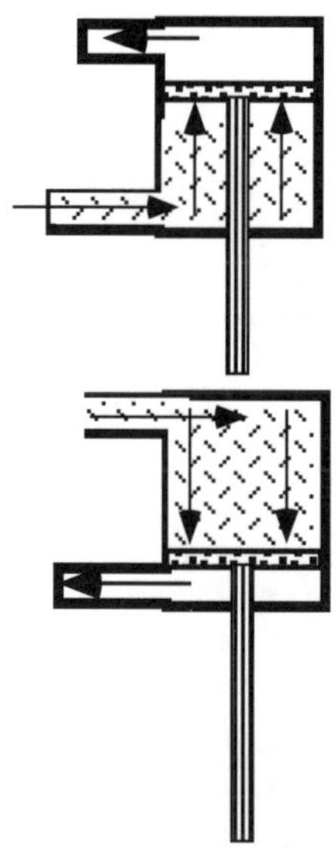

Double Action Steam used on both Sides

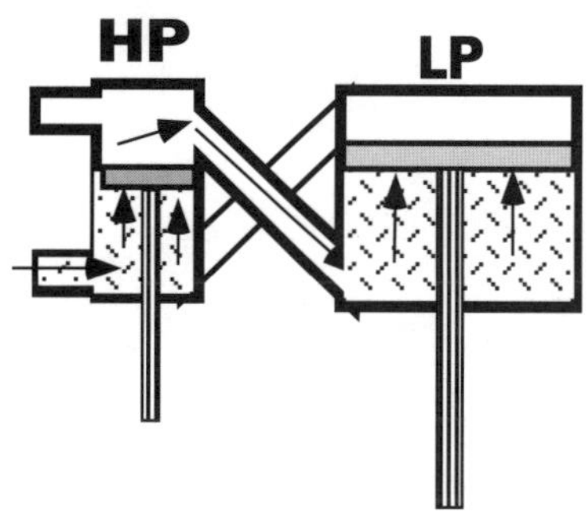

Compound Engine Steam expanded into larger cylinder

Triple Expansion Steam expanded 3 times

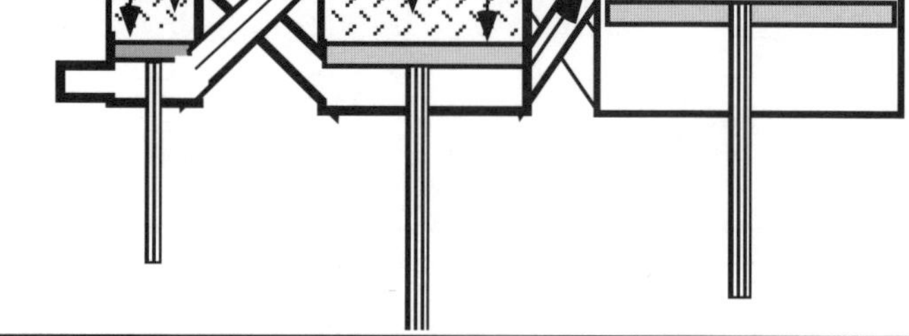

The terms compounding, triple expansion, quadruple expansion, single action and double action come up in literature on steam all the time. They all allude to how the steam acts on a cylinder and how many times it is used (expanded). A compound engine uses the steam twice. A triple expansion uses the steam three times. A quadruple uses it four times. You cannot tell the type of engine by counting cylinders as the steam may be expanded to more than one cylinder.

Condensers were another problem for early ocean-going vessels. Until the surface condenser was made workable the jet condenser was in general use. This injected cold sea water into the steam to cool it. The salt water then got back into the boiler with the feed water. The amount of salt water in the boiler could not be allowed to become excessive. The boilers had to be "blown down" to reduce the amount of salt water at regular intervals, further diminishing the efficiency of the boiler. This reduced the life of the boilers and added to the amount of down time for scaling. The effect of seawater on iron was not fully understood by many engineers. "The Use and Abuse of the Steam-Boiler" 1881, noted that "Marine boilers seldom last more than four or five years" and later on states "the difference (from land boilers) is not the effect of any chemical action of the sea water."

This was not as big a problem for river boats working in fresh water so jet condensers lasted longer in this trade. However most of the high pressure engines used in the river systems exhausted to atmosphere and did not use condensers.

The surface condenser is designed to pump sea water through tubes in contact with the steam cooling it. This lengthened the life of a boiler considerably. It also saved having to blow down the boilers.

The early engineers did not have the use of the modern lubricants. A list of them from a turn of the century engineering manual includes sperm oil, whale oil, lard oil, neats foot oil, linseed oil, cotton oil, palm oil, rape oil, castor oil, olive oil and the new mineral oils. Tallow being a favorite lubricant for years. Palm oil and tallow wound up mixed with steam in the condenser. This coated the tubes on surface condensers. Jet condensers were used longer than they should have been till this problem was solved. Tallow and other animal fats wound up in the boiler. It was discovered years later that they increased the deterioration of the iron.

The first engines used on screw (propeller) driven ships were converted from those used on paddlewheel ships. Gearing was employed to speed up the propeller shaft. This was not too successful. Some of the large gears had wooden teeth working with a metal pinion gear.

The direct-acting vertical engine was popular for early screw-driven vessels after some years of experimenting with horizontal and converted paddlewheel engines. By the 1860's there were a number of these engines used on smaller vessels.

In Europe from 1845-50 the screw was replacing the paddle wheel for some ocean-going ships. However the paddlewheel was still more viable in many areas on the West Coast. In Puget Sound and San Francisco harbor paddlewheel ships were still around in the early part of the 20th century. Rivers and lakes used them up till the 1950's. The last of the large wooden paddlers had pretty well disappeared by the 1870's. Some of them were brought back into service for the 1898 gold rush.

Small compound engine. The eccentrics on the shaft at the bottom. The linkage determines whether the ahead or astern link is used. The valve rod connected to the link operates the LP valve. Engine at San Francisco Maritime Historical Park Museum.

When boiler pressure became higher compounding, expanding the steam twice became feasible. Compound engines became common in the 1870's. Triple and quadruple-expansion engines followed in the 1880's, changing engine design till the end of the steam era. However there was not much confidence in them until around 1890. Some smaller coastal vessels made use of these engines before this. Screw-driven tugs using compound and later triple expansion engines started operating in B.C. and Puget Sound in the 1870's and 80's.

In newer sidewheelers a lot of multiple expansion inclined-acting engines were installed. Compounding oscillating were used till recent times in Europe. Sternwheelers generally were equipped with two horizontal direct-acting high pressure engines.

In England most battleships and cruisers in 1870 were still fitted with box boilers and horizontal engines. The compound engines were introduced to some naval vessels after this date. Triple expansion was introduced after 1880.

Later these were replaced by turbines in the faster ships. Turbines eventually replaced reciprocating engines on large ships in the years after WWII. Smaller vessels went to diesel.

The weight of engines and fuel was a constant problem in the design of steam ships. When pressures increased, the engines had to be built heavier until metals were improved.

There are basically two types of boilers in marine use, fire flue or tube and water tube. With the fire tube or flue boilers the gases and heat from combustion went through flues or tubes surrounded by water. With a water tube boiler the tubes are filled with water and the gas and heat is directed around them.

Boilers were always a problem until metals were improved. Boiler explosions were quite common in the early days especially in the rivers. Most of the early British boilers were rectangular in shape and were quite large. Most of the early U.S. built boilers were flue boilers of one type or another. They had a very low pressure. When cylindrical boilers were developed they were far more efficient. Some of these boilers were prone to blow up and the wooden hulls had a bad habit of going on fire. As you might imagine, burning or blowing up your passengers was not good for business.

Vertical water tube boilers were in general use in the U.S.N. in the 1860's. Some merchant vessels had them as well. They seem to have been used more in the U.S. than in Britain.

Boilers on later ships are of two types, "Scotch" or fire tube boilers and assorted types of water tube boilers. Most merchant ships, deep sea or coastal, were fitted with Scotch marine boilers up to WWII. Fast liners and warships had water tube boilers. River boats often had a type of flue tube boiler or railroad boiler. The flue boilers used on a lot of the Mississippi vessels were unlike the ones used on coastal and deep water ships. The water tube boiler with its higher pressures became the norm on larger coastal ships and deep sea ships after the war.

The heat is generated by firing coal or oil in the furnace. At one time wood was a common fuel for North American coasters and river boats. The hot air rising created the draft necessary for the fires to burn well, therefore, the tall funnels seen on these ships. The tall funnels kept a lot of the sparks from landing on deck and on paying passengers. The funnels got very hot, as much as 6000°F. In 1900 a typical Scotch boiler had a working pressure of 90 psi. In the 1960's it could have a working pressure of 220 psi.

A ship could have draught problems if the wind was blowing from the wrong angle making it difficult to keep steam pressure up, "Keeping it up to the blood." Later, forced draught, forcing air in the fires, helped this situation. The tubes got covered in soot and had to be blown out with steam. This was referred to as "punching the tubes" and if there was no wind, this could dump soot all over the deck. This caused a bit of animosity against the "black gang" from the deck crew that had to clean up the mess. The fires had to be cleaned at regular intervals as well. The fireman that could clean a fire without losing too much pressure usually escaped the wrath of the engineer.

Propellers

At first both paddlewheels and screws were called propellers. The sidewheel equipped ships had a lot of advantages at first but when engines, boilers and other machinery improved they were no longer useful except in rivers, lakes and inlets. The sternwheeler was proven to be far more useful than sidewheelers in the fast running narrow rivers of the West. At sea sidewheelers were prone to damage and the paddle boxes created resistance. The loaded draught of the ship created more problems. If the ship was too light the paddles would not dig in far enough. If she was to heavily loaded the paddles dug in too far. The rolling of the ship caused one paddle to be out of the water while the other was buried.

Screw-driven ships are not as affected by the rolling of the ship, her load, or debris in the water; though these factors make a difference to any ship. The early engines delayed the use of the screw-driven vessels. However they proved efficient in auxiliary sailing ships. The more efficient screw started taking over from the paddlewheelers in the late 1850's.

Part 2 - Steam

The theory and calculations involved in steam thermodynamics are very complicated. However a few basics will help the reader understand the operation of the engines.

Heat is applied to water to bring it to the boiling point where steam is formed. The heat from the steam does the work in the engine. Expanding and cooling steam releases this heat. Keeping this heat until it could be used by the engine was a major problem for engineers. For our purposes there are two types of steam.

Saturated steam-Steam that is in direct contact with the water it is formed from. It has the same temperature as the water. It is steam at the lowest temperature under which it can exist as a vapor. The temperature required to boil water increases as the pressure in the boiler increases. Steam produced in an ordinary boiler falls under this category.

Superheated steam-If a vessel containing saturated steam is heated and there is no water in the vessel, the temperature of the steam will rise. The steam will become supersaturated. This steam is dry and can stand a greater temperature drop without condensing in the cylinders. This saves fuel. It

A Steam Engine is a Heat Engine
1 Heat is released from the fuel when it is burnt. 2 The heat is put in the steam. 3 The heat is converted to work in the engine.

The application of heat to water.

When heat is applied to water the temperature is raised and the piston (A) will rise slightly due to the expansion of the water. When the water reaches the boiling point evaporation starts and the piston starts to rise (B). As more water is evaporated the piston will rise higher (C). Steam in contact with water is saturated steam. It has the same temperature as the water. This is the boiling point under the available pressure. To increase the temperature you must increase the pressure in the container.

If saturated steam is taken from the container with the water and heated in another container (D) it is superheated steam. The temperature can be raised and the pressure increased accordingly. The temperature is increased without an increase in weight.

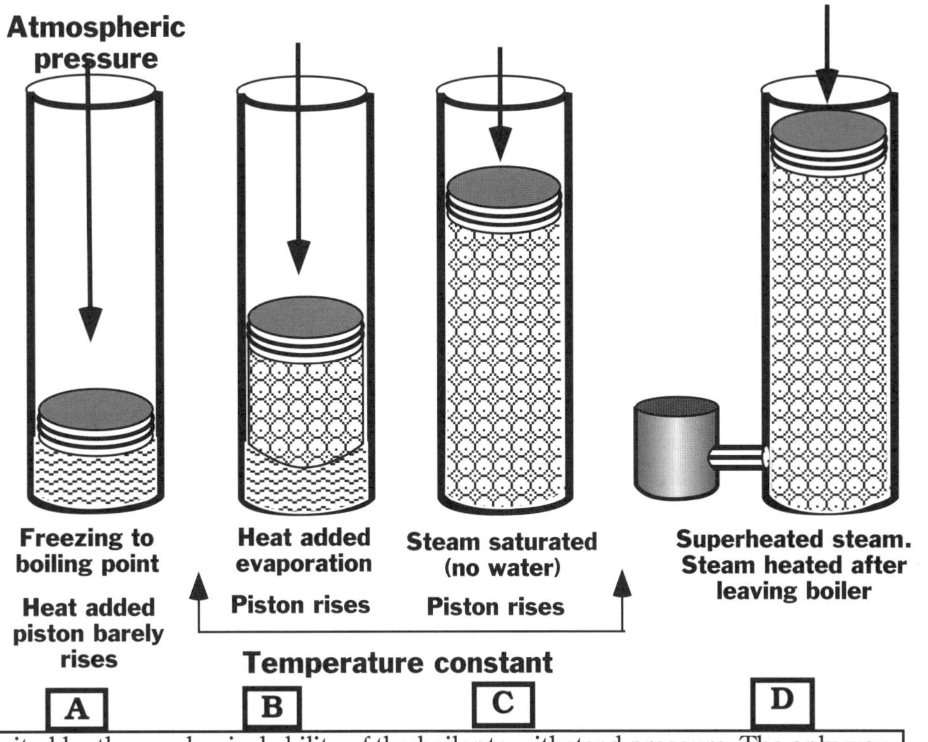

This means that steam pressures are limited by the mechanical ability of the boiler to withstand pressure. The only way early engineers could get higher pressures to make the engines more fuel efficient was to increase boiler pressure, which created the odd boiler explosion, or use superheated steam. Today's turbines need superheated steam from a high pressure boiler to be effective.

is nearly always used with turbines. It is not as good for older type reciprocating engines. It is usually created by passing steam through heated tubes. The U.S. Navy experimented with superheated steam as far back as the 1860's in an effort to reduce condensation in the big low pressure engines.

1828 saw the first known use of superheated steam. Its use dropped off from 1870 to 1900 but became more popular later.

One of the problems in designing steam engines is trying to get the most energy out of the least amount of fuel. As metals improved, greater pressures could be employed making the engines more fuel efficient. However there are limits to what could be done and modern diesels are far better in this regard. The other problem is space. By comparison, a boiler, engine, and water tanks take up far more valuable space needed for cargo or passengers, than a diesel engine of the same horse power. This is especially true on tugs and other small coasters.

It was not only the machinery that had to be developed on the early steamships. Ship construction had to undergo some major changes. The added weights and vibration caused major problems on wooden hulls. The smaller wooden coasters worked out well but the larger ocean-going vessels had problems. Wooden ships work in a sea. The hogging and sagging in a sea was a real problem for larger steamers. Engine parts and boilers needed to be ridged. The lack of stiffness in the hulls slowed down the development of screw-driven ships. Composite hulls, with iron diagonal strapping and knees became necessary. It was not till iron hulls and the yards to build them were developed that a lot of these problems were reduced. Europe was ahead of North America in the development of these hulls. In the U.K small iron hulled vessels were built in the 1830's. The Great Britain was the first large ocean-going iron hull built in 1843. Due to Admiralty requirements larger iron-hulled steam ships were not built until the 1840's. In the U.S. small iron hulls were built in the 1840's but the first larger iron steam ship was the schooner rigged Champion built in 1859.

However, wooden steam tugs over 200 ft.(61m) and freighters 280 ft.(85m) were built in WWII. The tugs, usually converted to diesel, operated for years after the war.

Engine Facts

When reading about steam equipment the reader should understand that most of the engines were usually individually made and consequently were quite distinctive in design. The exception to this was during wartime. Steam engines were produced in large numbers from the same plans. The names of engine types can be misleading as changes in names were made over the years or area. Remember the boiler pressure and type of steam used today is not what was used when the first steam ships sailed. Before compounding engines each cylinder was referred to as a separate engine even if they were on the same crank shaft. This terminology was used at times even after compound and triple expansion engines were in existence.

Types of Valves

All engines need some type of valves to introduce the steam pressure to one side of the piston and let the steam exhaust from the other side of the piston. There have been many types of engine valves used over the years. The three basic types are poppet valves, slide valves and piston valves.

Balanced Poppet Valves

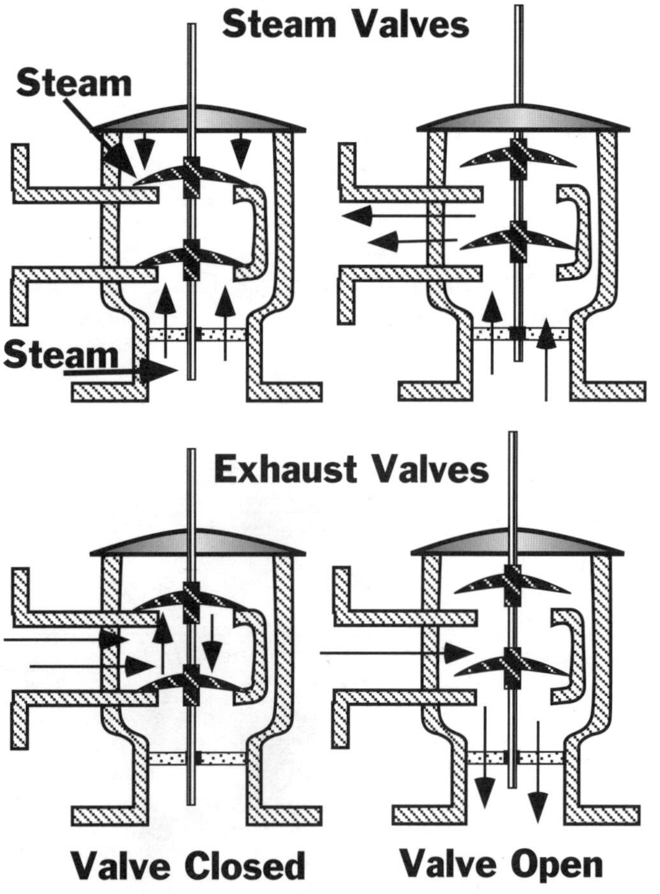

The Poppet Valve

The poppet valve has been used for years. On higher pressures it can leak. The single poppet valve was hard to operate under higher pressures as it had to overcome the steam pressure to open. The double poppet valve overcame this problem by having steam on both sides (balanced).

Two sets of valves had to be used in each end of the cylinder, a steam valve and an exhaust valve. When used on the steam side (top illustration) the valve was lifted to let steam pressure into the cylinder. On the other end of the cylinder the exhaust valve had to be opened.

In the early days the U.S. had great success with this valve while the Europeans had problems. Later it was used in some applications all over.

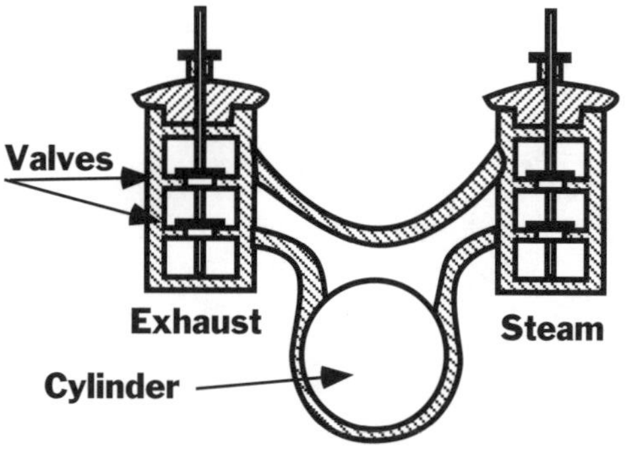

Poppet valves on a horizontal engine. Most of the West Coast river and lake sternwheelers used slide or piston valves.

Steam propelling equipment usually consists of a boiler, condenser, engine and auxiliary equipment. Steam was also used to supply power on deck to winches, pumps and heating. Some later sailing ships used steam winches to do a lot of the heavy work so the size of the crew could be reduced.

With all types of steam engines, water is heated in a boiler till steam is created, stored in a steam chest and delivered to the engine as required. After the steam is used in the engine it goes to the condenser in most cases. It then goes to a hotwell before being put back in the boiler as feedwater. On newer vessels with high pressure boilers, the water will be further heated before going back to the boiler. Some engines such as winch engines and some high pressure paddlewheel engines did not have condensers. These engines are said to "exhaust to atmosphere."

If the steam is used only on one side of the piston it is a single-acting engine. If steam is used on both sides of the piston it is double acting. If the steam from the first cylinder (high pressure) is used again in the second cylinder (low pressure) it is a compound. If the steam is used again in a third

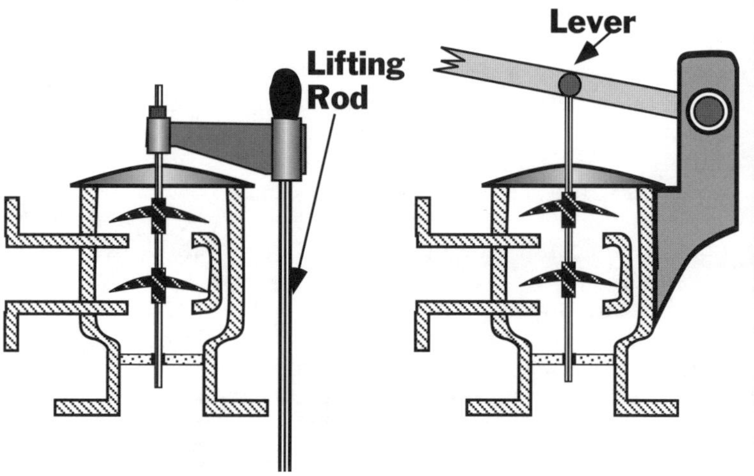

The poppet valve could be opened by a lifting rod, such as used on the walking beam engine, a cam, or a lever similar to the type used on early river boats.

These valves were used for throttle valves on some engines.

The Slide Valve

The slide valve is the most common valve used on piston type engines. They were used on the older engines as well as the newer reciprocating engines. There are many versions of this valve in use. With very high pressures there could be some problems so they might not be used on the high pressure cylinder. Some of them had several ports. Of course turbines are the one engine where they are not used.

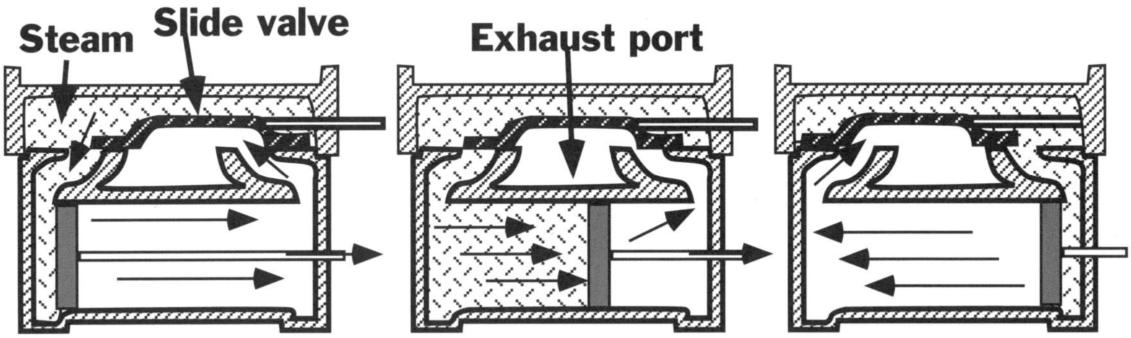

Steam enters left exhausts right.

Valve cuts off. Expanding steam keeps piston moving.

Steam enters right exhausts left.

The slide valve as the name implies, slides back and forth letting steam in one end of the cylinder while exhausting steam out of the other. On faster engines it can be set to cut off the steam before the piston reaches the end of the cylinder, letting the expanding steam do the work. The D valve shown was used as a locomotive valve. There could be a problem with variations in pressure between the front and the back of the valve if a relief frame wasn't used.

Lap - Amount outer edge overlaps ports when valve is in mean (center) position.
Lead - Amount valve is open at the start of the piston's stroke.This allows a cushion of steam at the end of the stroke. Some engineers allow more lead at the bottom of the stroke on a vertical engine.

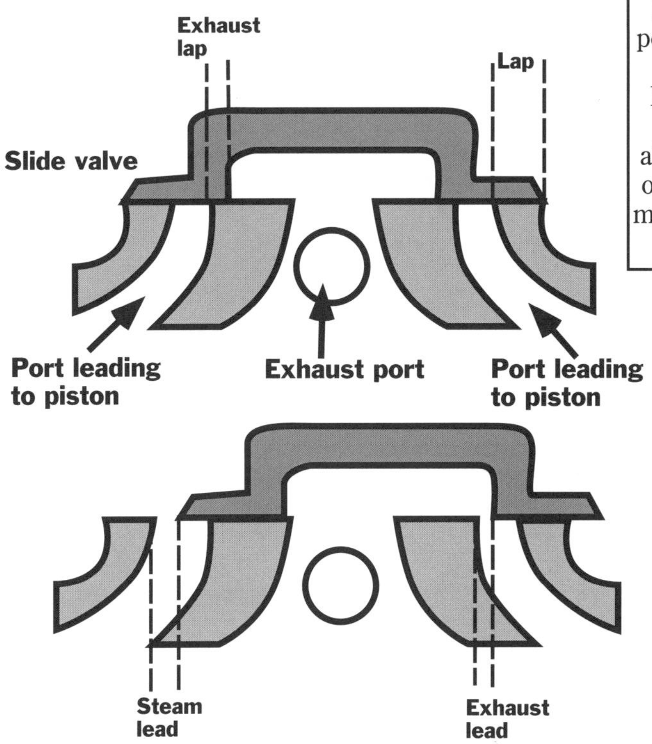

Slide valve terms.
A simple D slide valve and the cylinder ports.

cylinder, HP (high pressure) to IM (intermediate) to LP (low pressure), it is a triple expansion. The most elementary form of double-acting engine has only one cylinder. The steam would be admitted to each end of the cylinder alternatively and the exhausted steam expelled to the atmosphere.

In order to induce and exhaust steam at the right time for the engine to operate, a steam box and valves must be used. The opening and closing of these valves was predominantly done with rods connected to an eccentric on the crank shaft. There were two types of eccentrics used, loose and fixed. The fixed system used a separate eccentric bearing for ahead and astern while the loose system used only one. Most of the early engines used a loose eccentric. Unfortunately the stop used to carry away now and then.

The valves used in engines were derived from three basic types.

1 - Slide valves
2 - Piston valves
3 - Poppet valves

Separate cut-off valves were used on some engines. These valves cut off the steam to the cylinder before the end of the stroke. This allowed the expanding steam to continue working. This reduced the amount of steam and fuel needed. On many engines this can be accomplished with the slide valve.

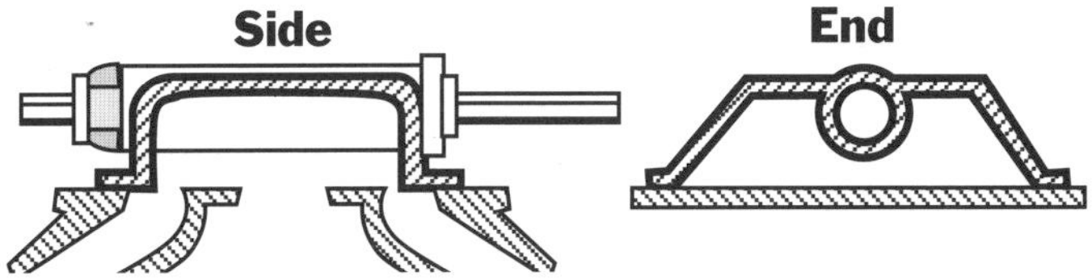

The Piston Valve

The piston valve worked well under higher pressures. As with all valve types there are many versions around.

The Corliss Valve

Corliss valves were used in a number of stationary engines ashore. Although they were not used too much at sea some marine engines and riverboat engines had them. Unlike other valves they revolved. As the steam valve could be operated separately from the exhaust it was easier to adjust the cut off. There were several versions around including a flat type. Economy of steam was a major advantage. The steam entered the cylinder just under boiler pressure. Steam could be cut-off quickly letting the steam expand until it could do no useful work.

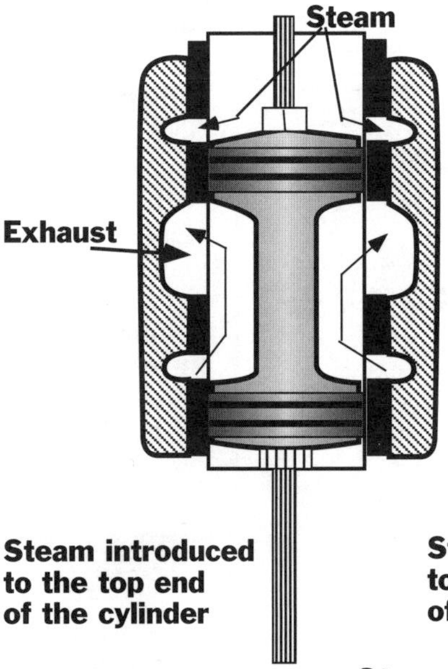

Steam introduced to the top end of the cylinder

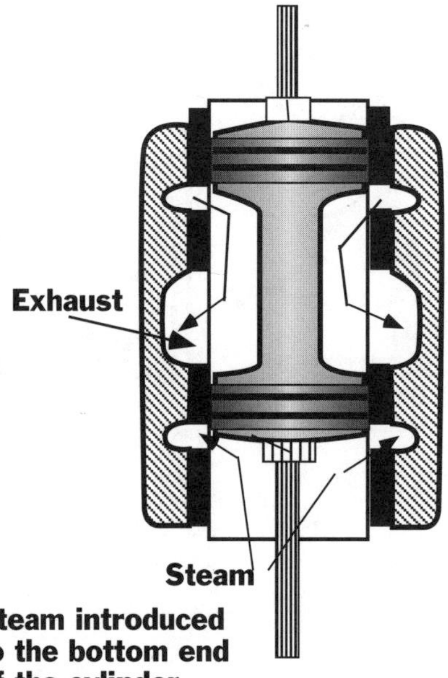

Steam introduced to the bottom end of the cylinder

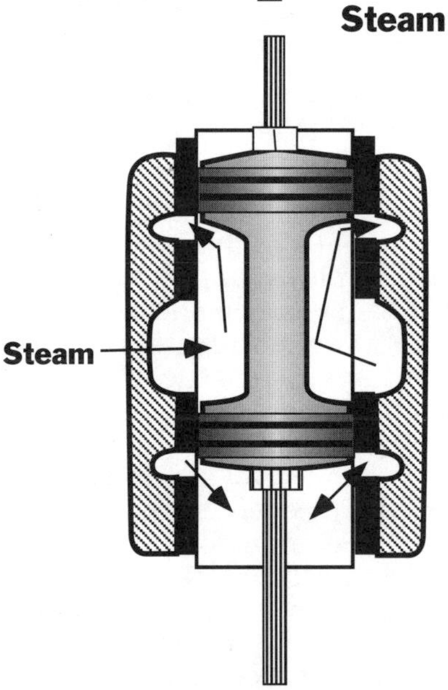

Steam inside

The piston valve taking the steam from the center offers the following advantages: easy to keep tight.

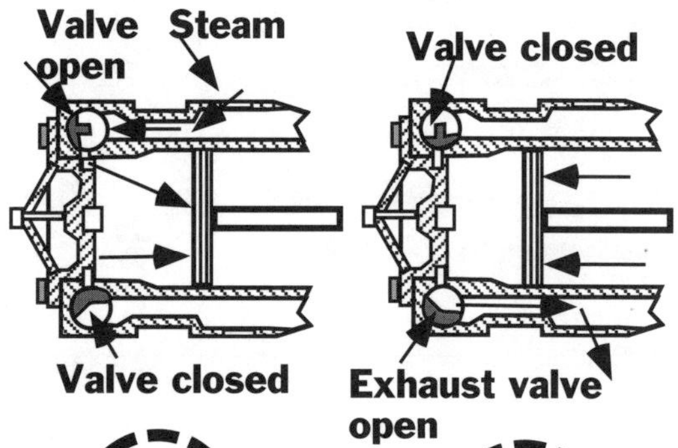

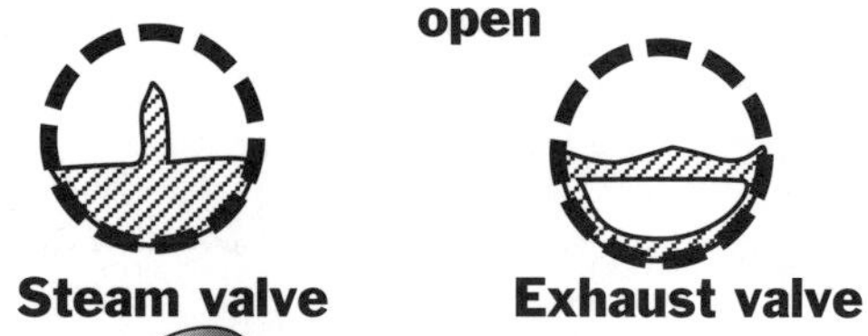

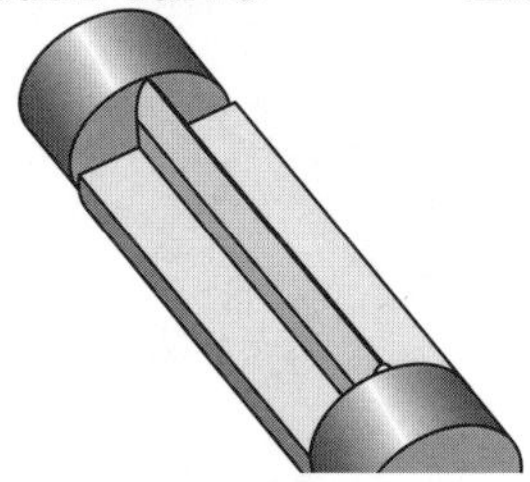

A piston valve is actually two pistons on one spindle. It works in a cylinder with ports running diagonally around the chamber. The pistons cover these ports with around the same amount of lap as a slide valve.
The steam can be admitted from outside the piston or inside. It is frequently used for the HP cylinder.
The advantage of these valves: greater port area than a flat valve, a balanced valve for higher pressures, less friction than an ordinary slide valve.
Used early in the 19th century.

To obtain maximum efficiency from a piston type engine several things must occur:

1. There must be a vacuum or lower pressure on the side of the piston where the steam is being exhausted.
2. There must be sufficient steam pressure.
3. The highest fuel efficiency is obtained by using high pressures.
4. The steam entering the cylinder must be cut off to let the steam expand in the cylinder.
5. With higher pressures stage expansion is necessary to avoid condensation. This is around 70 psi for compounding, 120 psi for triple and 215 psi for quadruple.
6. The cylinder must be kept as warm as possible (not too much expansion in one stage).
7. Condensation of steam on the colder cylinder walls with the resulting loss of efficiency must be reduced as much as possible.
8. The condensed steam must not cool too much before it is injected back into the boiler as feed water.
9. The direction the engine turns, ahead or

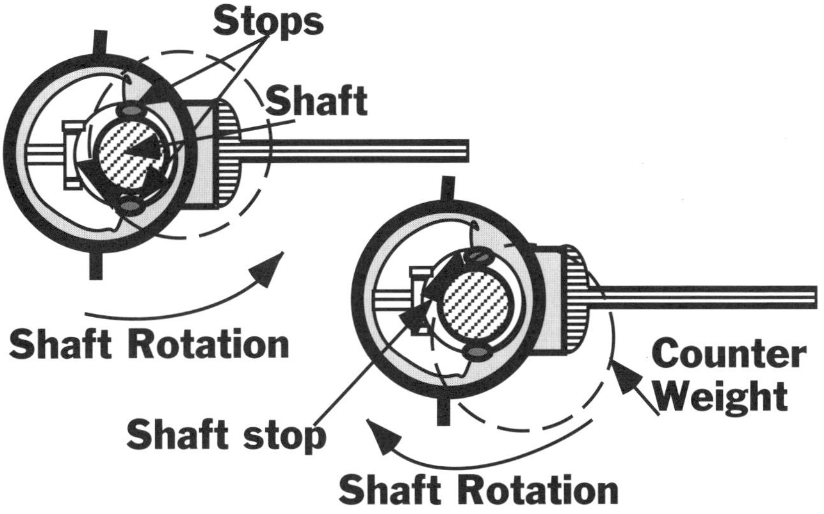

Loose eccentric - Paddle wheel
After getting the engine turning over with the use of starting lever the eccentric is hooked on the stop and the shaft engages the stop on the eccentric. The valves are operated in the correct sequence. When reversing the other end of the shaft stop engages the other eccentric stop.
A counter weight is used for balance.

The valves of all types were most often operated by an eccentric bearing on the main crank shaft. Timing the valves to open and close at the right time was critical. The engineer had to adjust them to allow for lap etc. The eccentric works as a small crank to move the valve or valves the right distance.
At **A** with the piston at one end of the cylinder the eccentric starts to open the valve.
At **B** the valve is fully open as the piston moves in the cylinder. The steam is exhausted from the other side of the cylinder.
At **C** the valve is reversing its direction letting steam into the other end of the cylinder.
At **D** the valve is wide open allowing the full pressure of the steam to the piston.

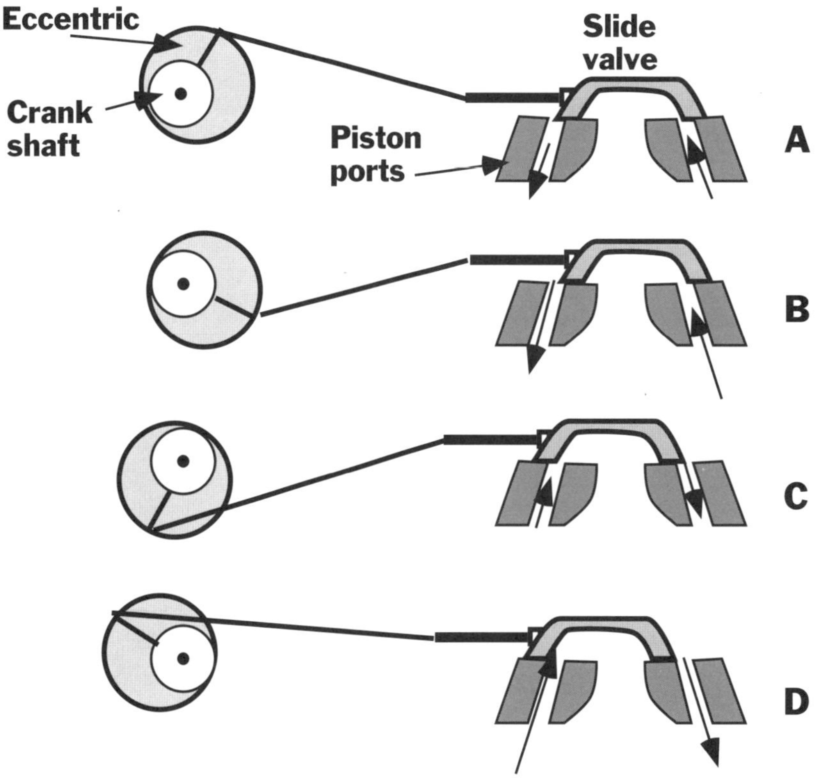

The eccentric bearings on the crank shaft. They can be adjusted. The eccentric straps fitted over them and were free to move as the shaft rotated.

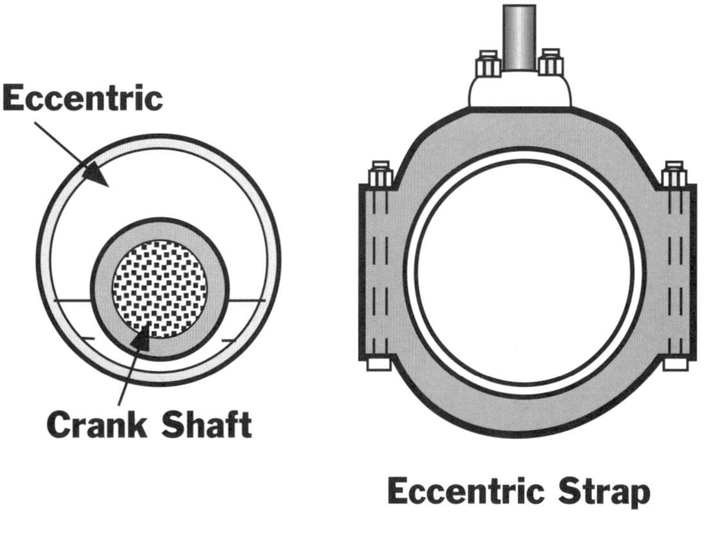

The revolving crankshaft causes the eccentric to rotate. The eccentric strap works outside of it. The motion of the eccentric makes the eccentric rod reciprocate moving the valve.

astern, is regulated by the valves letting steam into the cylinder and exhausting it. The valves are operated by eccentrics on the shaft.

10. Rotary engines (turbines) with a good condenser are more efficient than piston engines.

The steam engine was a very reliable engine with the proper care. Tolerances were far greater than internal combustion engines so the engineer could do a lot of the work on the engine and its bearings in any port. A lot of the parts that did not require large castings could be made or repaired in any good marine machine shop. Unfortunately when comparing them to a modern diesel they were heavy, not fuel efficient, took up a lot of space and needed a bigger crew. But without them, world trade as we know it would not have developed.

Horsepower

Unit of power equal to that required to lift 33,000 lb. one foot in one minute. (550 ft-lbs per second).

There have been many different ways of calculating HP. This can create a bit of confusion.

Nominal horsepower NHP - Used in the "Blue Book" and government regulations.

Survived from the early days of steam. The boiler pressure and piston speed were constant. Gives a much lower number than other horse power ratings. For years engine makers used this as it is independent of the steam pressure provided by the boilers. There have been many ways of calculating it. Nominal horse power is almost useless.

Indicated hp IHP - Used on most steam engines. Indicator diagrams are made from pressures taken on the individual cylinder by an indicator instrument. This was done frequently on large ships.

Shaft horsepower SHP - As indicator cards cannot be taken on turbines shaft horsepower is taken with a torsion meter.

Brake horsepower BHP - The actual horse power put out by an engine. There are a number of methods used to determine the force.

Static bollard pull SHP - Used to measure the amount of force a tug is capable of applying to a tow under certain conditions.

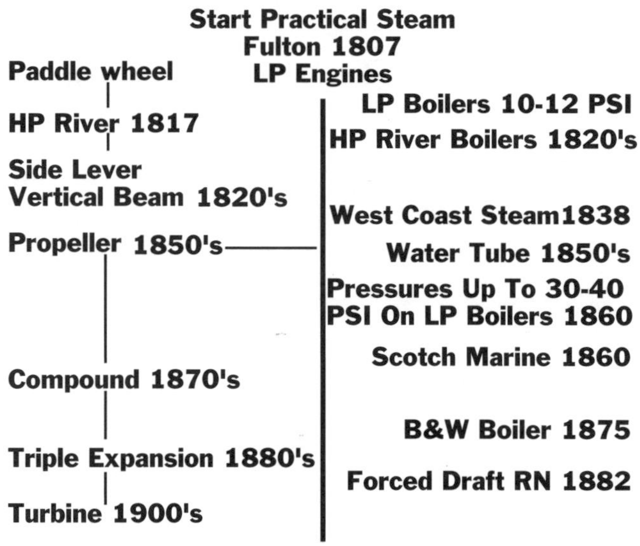

Approximate chronological order of steam engine development.Steam engines and metallurgy did not develop at the same time in all parts of the world.

Boilers
Chapter 3

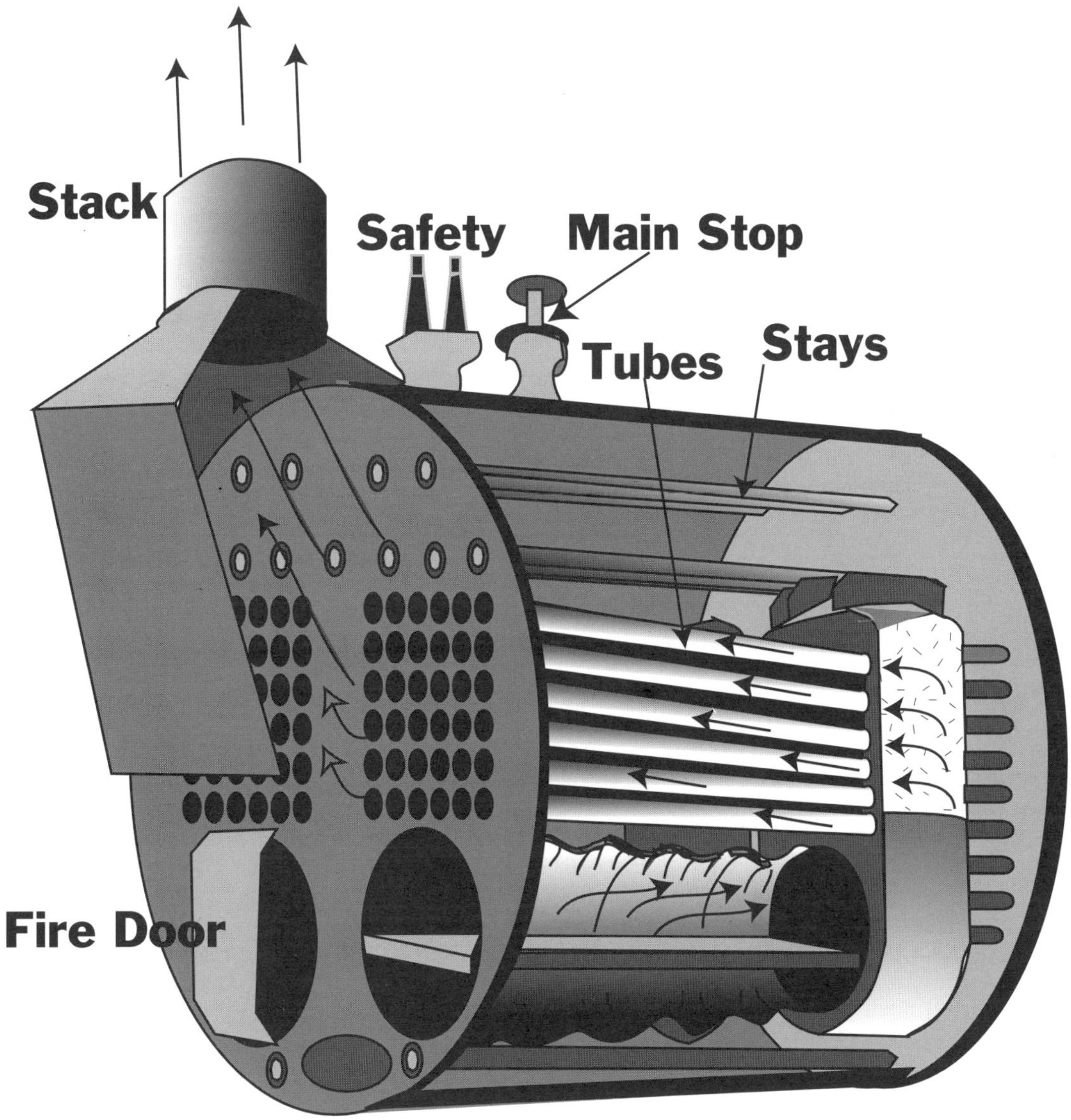

Scotch Boiler

The scotch marine was probably the most common type of boiler used on the West Coast. It was a very reliable boiler and relatively easy to fire and maintain. It was the mainstay of the towing industry though there were some water tube boilers used.

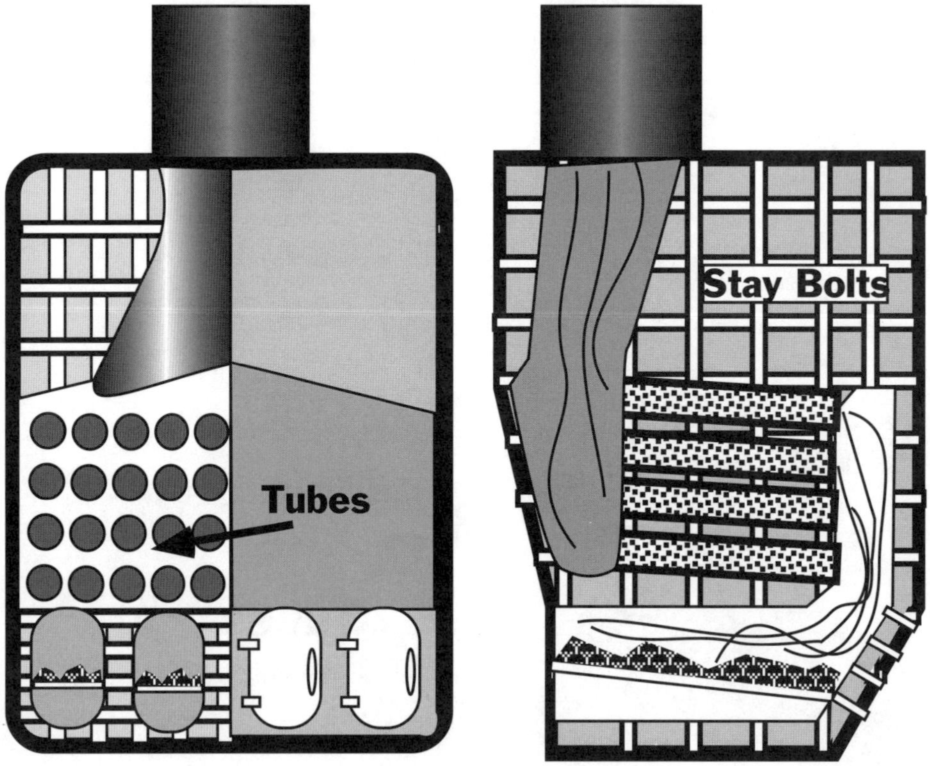

Box (Haystack) Boiler 1870's. Working pressure 30 psi.
The first box boilers were very low pressure operating at around 10 psi. This type of boiler depended on stays (metal rods) running in all four directions to keep their shape. The stays made them difficult to clean. These boilers were more common in Europe than in North America.

Boilers

Regardless of the type of engine a steam vessel may have, she must have a reliable supply of steam from a boiler. On larger ships unlike small coasters in addition to the "main steam boiler or boilers", a vessel may have an auxiliary boiler or a donkey boiler. The auxiliary boiler can be used for purposes such as heat and electricity. This is advantageous in port when the main boilers are on standby. Even modern diesel-powered tankers often have auxiliary boilers aboard. A donkey boiler is a steam boiler installed on a vessel for occasional use. Later day sailing ships sometimes had them aboard. Shipowners could reduce the crew size by using steam for the heavy lifts. They were often vertical-type boilers.

There were many types of boilers used on the West Coast. As some of the ships or their machinery originated elsewhere a description of the history of boilers is in order. A ship may have had assorted types of boilers in its career because unlike the engines, the older boilers did not have a long life.

To be efficient a boiler should produce the greatest amount of steam for the least amount of fuel. This has never been an easy task. The lack of good material for the boilers was a major problem for early engineering companies. Good construction methods took years to develop. In addition the demands of the trade a vessel was engaged in created diverse boiler requirements.

In an effort to overcome the many problems encountered with boilers, there have been enough designs to fill an encyclopedia. They can all be separated into one of two categories, fire tube/flue or water-tube boilers. With the fire-tube boilers the combustion gases and heat from the furnaces were drawn through the tubes and up through the stack heating the surrounding water. With a water-tube boiler the tubes were filled with water and were heated by the combustion gases moving past them.

Boilers were either externally or internally fired. In a lot of early boilers, the furnace was made of brick and the actual boiler was mounted on top of the furnace. This practice came from the boilers used in stationary plants ashore. Some of them had a single drum and an outer shell curved on the top with flat sides and bottom, known as a wagon top boiler. Others of this type had several drums and furnaces. They did not have the steam drums or domes. If the fire was inside the boiler it was called an internally fired or firebox boiler.

As long as jet condensers where used on vessels sailing on salt water, a real problem existed due to the amount of salt water injected into the boilers. Boilers used in salt water using jet condensers had to be shut down and scaled frequently. Flues often needed replacement. They frequently had to be blown down with fresh salt water to reduce the salt content in the boiler while at sea. This had to be every two hours on flue boilers and every 20 minutes on tubular boilers. In addition a lot of time was lost for maintenance while in port.

The types of boilers have changed over the years. The first boilers were large copper boilers

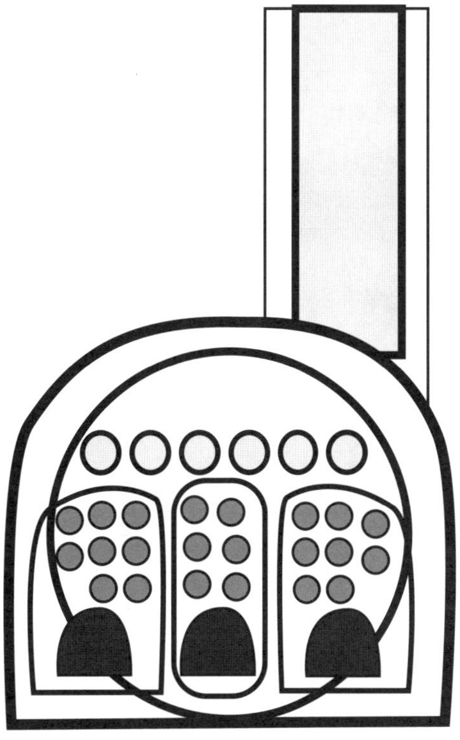

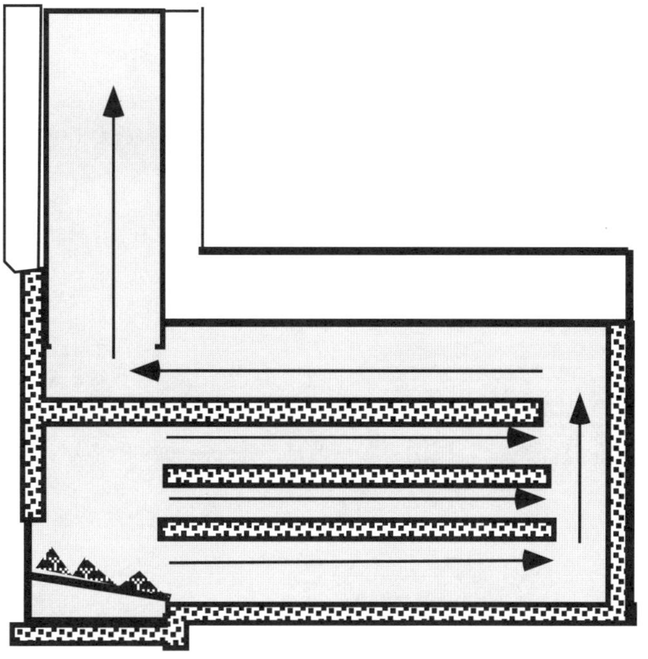

The return flue boiler
One type of early low pressure internally fired boiler. The combustion gases and smoke follow the flues to the back of the boiler where they are returned to the front of the boiler by a large flue before going up the stack.
These are schematic type diagrams. They do not show plates and combustion chamber

operating just over atmospheric pressure. The first practical European ones were generally rectangular in shape and had a very low operating pressure (around 10 psi or lower). The maximum pressure this type of boiler could handle was 30-40 psi. The box type boiler fitted the shape of the ship better than some of the other designs. The pressure was limited as they depended on iron stays for their strength.

In addition to some box type boilers a lot of the early American boilers were cylindrical. When metals and design improved, the pressure increased. The first engines were not very efficient. They used an awful lot of steam and fuel. Most of the older ships had boilers with many furnaces side by side. A large crew was kept busy keeping the fires supplied with fuel. They worked in the heat and relative darkness and often in a rolling ship. There were no electric lights, only oil lamps on the old ships. The coal passers lacked ventilation in the coal dust-laden bunkers.

At one time there was a tendency for boilers to blow up. There were a number of theories as to why boilers would blow up. Some of the theories would make any modern engineer laugh. One of these according to an 1880 manual was explosive gases formed by decomposing steam. Looking through old manuals makes fascinating reading.

It took some time for the government to catch up with the new technology. The first steamboat regulations in the U.S. came into effect in 1838. The story is as the inspectors only got paid $5.00 per inspection they just visually inspected the hull and machinery and never condemned any of the ships. The amount of pressure carried was up to the engineer and the master. In 1852, after some disasters, new laws were brought in requiring certificates for the engineer, the master and inspection of the boilers. In Britain the Merchant Shipping Act was enacted in 1854 consolidating and improving existing regulations. Compulsory examinations for

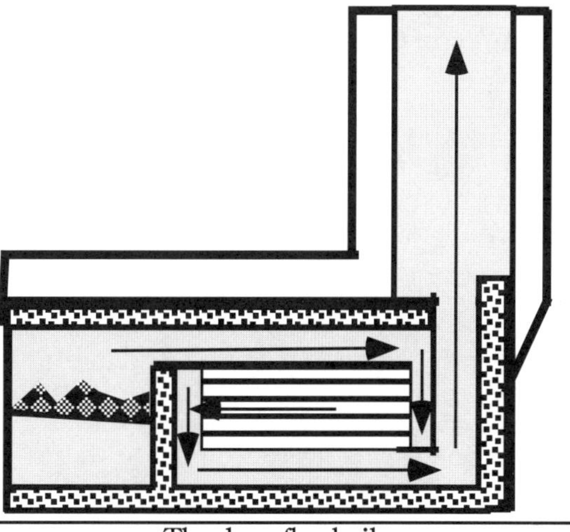

The drop flue boiler
Another type of early low pressure boiler. The gases went back and forth through the flues before going up the stack. On most ships a number of them would have been placed side by side.

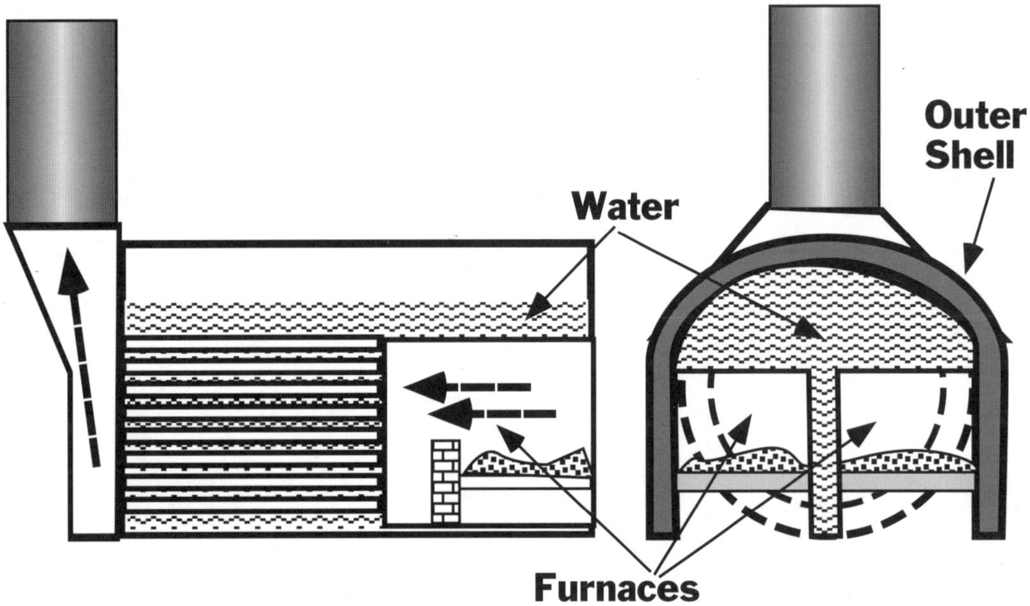

The tubular boiler was not as common as the flue boiler. They were both low pressure boilers. The tubular boiler was harder to clean and maintain. Note the outer shell. The Del Norte, wrecked off Porlier Pass, had two boilers of similar design.

deck officers came into effect in 1850 and engineers in 1861. Before this, standards were set by Lloyds. Enforcing these regulations on the West Coast was another problem for some years. These regulations were brought in for the sake of passengers and cargo. Crews were expendable.

Undoubtedly part of the reason various early boilers exploded was lack of knowledge on the part of the engine room crew. A lot of the problem was in the boilers themselves. Inspection and upkeep was not too efficient. One manual written in the 1850's suggests putting a pressure gauge on each boiler.

The quality control of the metal used in early boilers was somewhat unreliable. At one time North America was behind Europe in this respect. Wrought iron does not crack as easily as cast iron but it was hard to produce in sizes large enough for end plates. Slag and other impurities were a major problem for manufacturers. The advent of steel boilers overcame a lot of these problems.

On a lot of early vessels the feed water was fed to the boiler by a pump connected to the engine. If the engine was stopped, the feed water had to be pumped by hand. When the engine started again, cold water could be introduced to the boiler. This sometimes caused ruptures if the water level was low. Eventually separate feed water pumps and heaters were introduced.

It seems there were far more boiler explosions on river boats than on ocean-going vessels. This may have been due in part to the design and poor construction of the boilers. Shallow draft river boats required longer thinner boilers. The water level in the boiler was far more critical than in the larger boilers used on ocean-going ships. If a flue was not covered or cold feed water was injected it would blow up. Some of the old riverboats did not have pressure gauges. Some sources say they did not have water gauges or any method of keeping track of the water level but they would have had Gage (test) cocks. "The Use And Abuse Of The Steam Boiler"(1881) states that gage-cocks are more reliable but that the "glass water-gage" had been used almost universally since its introduction. It also mentions that the equipment was not well maintained. The safety valves were of the lever type which even if not tampered with could leak or stick.

One of the other troubles that river boats had was foaming. The feed water taken from the river was often muddy. This, along with boiler forcing (trying to produce more steam than the boiler was designed for) caused the surface of the boiler water to foam. The foam could give a false reading in the glass or gage cocks. In later years a gage operated by a float on the water was installed. This float was not affected by foaming. Early European manuals mention a similar device.

The last boiler off the Beaver. The first one was a square type. The boiler would sit over an external furnace. There was a large steam drum on top of the boiler.

When the boiler blew it usually spread the housework, people and boiler parts all over the place. On sternwheelers the wheelhouse was over the boiler so if the boiler blew the skipper or pilot usually was killed. A number of these vessels were rebuilt which is more than could be said for the crew or passengers. To avoid losing paying first-class passengers some owners tried putting the boilers outside the main housework, on the guards.

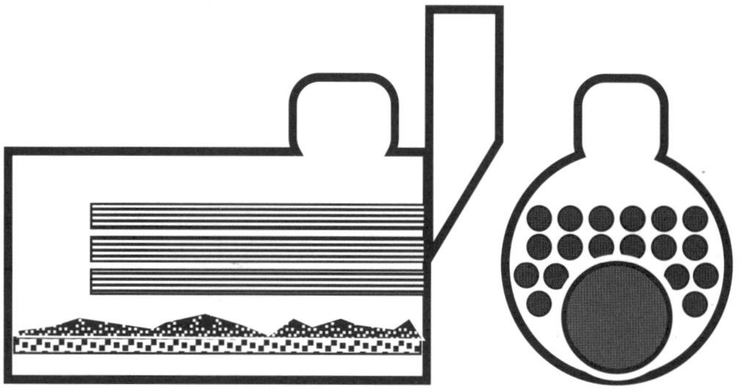

A single fire boiler common on smaller craft. Small boats and launches also used vertical boilers a lot.

Fire Tube-Flue Boilers

The difference between a tube and a flue is the diameter. Flues are far bigger than tubes. There were some large flues that were not round on early English boilers.

Heating the water efficiently is a matter of compromise in boiler design. In theory combustion gases should be in contact with the heating surfaces in tubes or flues until they become cooled enough that the transmission of heat stops (600-800°F 315-427C). In a cylindrical boiler the furnaces form the most efficient heating surface followed by the tops of the combustion chambers. The tubes come last. The top layers of the tubes do more heating than the lower layer. The first few inches of a tube is the most efficient part of the tube. Some engineers considered flues easier to clean than tubes. Soot, ashes, cinders, and salt from leakage tended to plug the tubes. However the smaller tubes were more efficient.

The haystack boiler was an early form of box fire tubular boiler. It could be dry bottomed or wet bottomed. The wet bottomed version had the furnaces including the bottom surrounded by water. The dry bottomed had just steel plates to catch the ash under the fire. The Royal Navy preferred the wet bottom type. They depended on stays running in all directions to hold them together. There were still some of them around at the turn of the century.

Another type of early boiler was known as a leg boiler. It used large flues to send the combustion gases back to a combustion chamber.

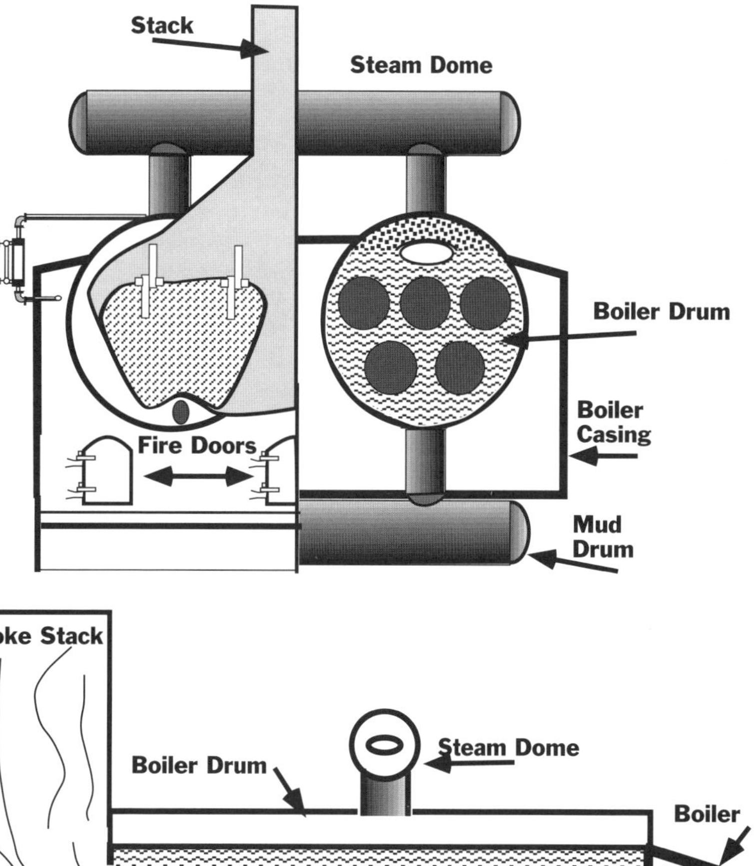

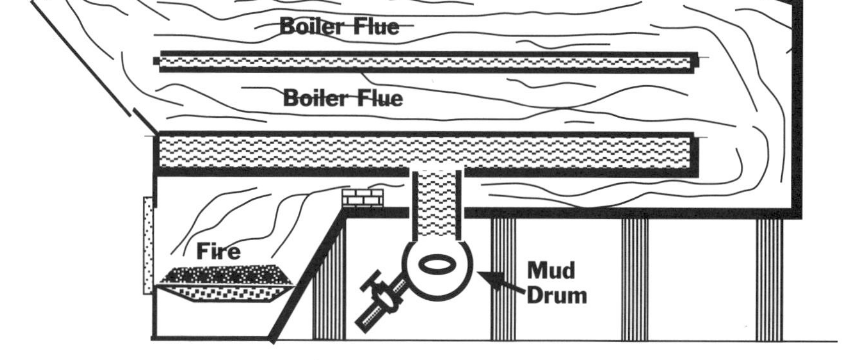

The type of externally fired high pressure boiler used on the Mississippi river boats. Not to scale. There were some used on the West Coast. There were usually more boilers placed side by side than shown and they were longer. Normally they used two stacks. The combustion gases went under the boiler and back through the flues. The mud drum collected residue and as it was connected to all the boilers feed water could be injected to all the boilers. Some boilers had two mud drums. Steam was collected in the steam dome.

Boiler cocks and water glass. The fireman could tell how much water there was in the boiler by opening each cock in turn. If steam came out the bottom one he would have to get water in the boiler before it blew.
Boiler on sternwheeler W.T.Preston.

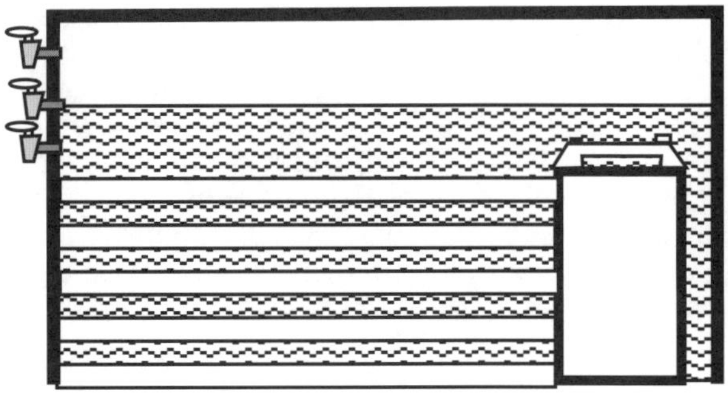

The bottom cock was 3 or 4 in. (7.5 to 10 cm.) above the crown sheet or heating surface. The center one was at the normal water level and the top one in the steam area. It required some experience to get a correct reading especially if there was foaming on the surface or if the boiler was forced.

Smaller tubes took the hot gases and smoke to the front of the boiler and up the stack. Some had a "steam chimney" to super-heat the steam. They were commonly used for tugs and river boats.

One of the earliest types of boiler was the Flue boiler. Iron return flue and drop flue boilers were used on most American ships in the 1840's - 1860's. A variety of flue boilers were used on western river boats for years after they were discontinued on larger ships. They could be found on some West Coast river boats as well. There were several types of these boilers built. The return flue boiler was probably the most common. Another similar type boiler in use on early steamers was called a drop-flue boiler. The combustion gases went down through the flues before going up through the stack.

There was also a return tube boiler in use where tubes were used instead of the much larger flues. Often two or more flue-type boilers were installed side by side with the smoke and gases going up one stack. Some boilers used flues and tubes.

A number of small boilers of horizontal return-tubular and flue-type had a dome attached at the top of the boiler shell. The steam dome is used to increase steam space and to obtain drier steam and prevent priming. The steam is drier because it is farther away from the water. The dome shell is usually flanged and riveted to the boiler shell. A flanged head closes the dome at the top. It is reinforced with stays or brackets. The hole or

A triple pass boiler on the sternwheeler Samson V. The gases go through the tubes three times before going up the stack. She was an oil burner.

The pumps and heaters required for the burners on the boiler. The oil was heated to the correct temperature and pumped to the burner where it was sprayed into the furnace for combustion. Samson V

The locomotive type boiler used on many sternwheelers. They had a low center of gravity which was a necessity on these shallow draft ships. They could get up steam very quickly but they primed easily (picked up water with the steam). On these boats the exhausted steam was run back to the smoke box. The combination of the steam and the smoke going up the stack created a good draught for the fires. This was technically called "blasting". With the exception of some smaller warships they were not used on ocean-going ships.

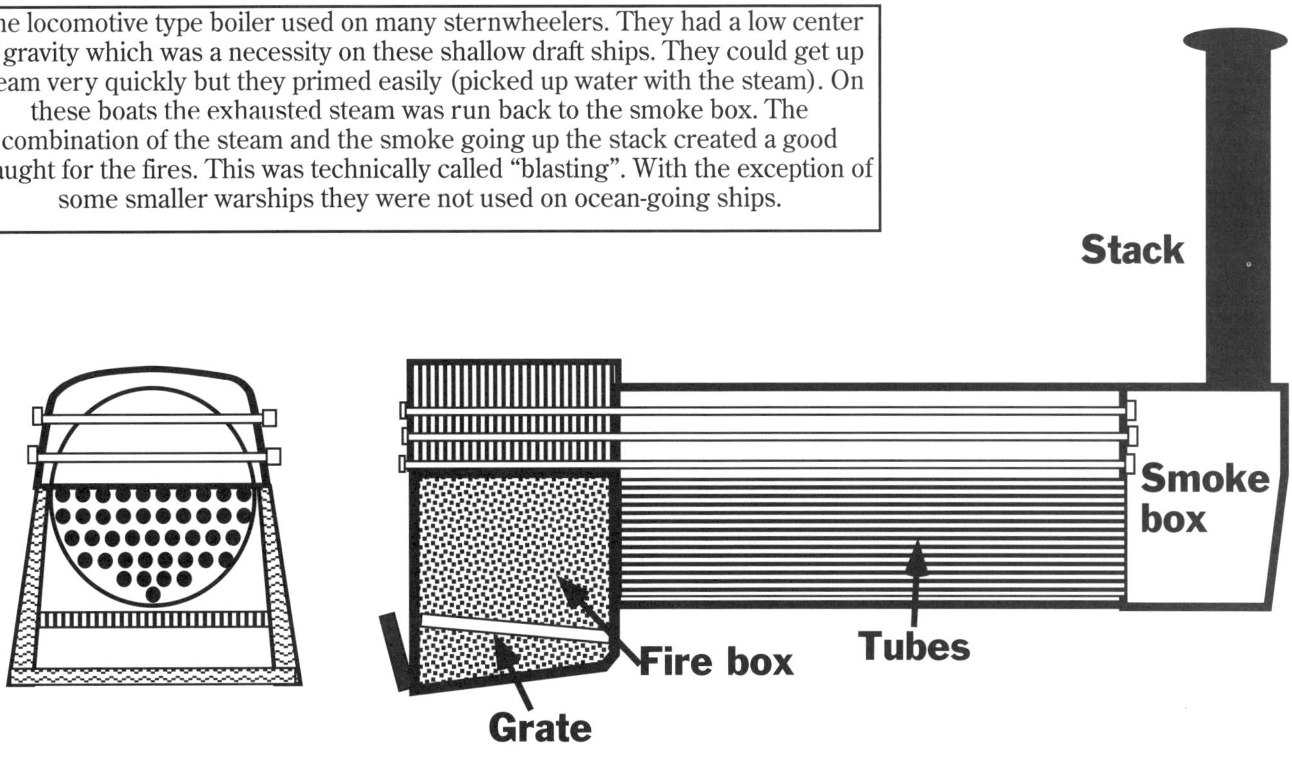

holes cut in the boiler shell to admit steam must be bigger than the area of the steam outlet.

Dry pipes and dry boxes were used in later years for the same purpose. They were often fitted in the steam drum of a water tube boiler. This prevents water being carried with the steam to the engines or cause water hammer in the pipes. This is known as priming. Pipes and valves can break or engine damage such as blowing out cylinder heads can occur.

River Boats

The very first river boats used on the West Coast were western rivers (Mississippi) sidewheelers. They proved impractical and after a few years a distinctive western-type sternwheeler was developed. Some western rivers' boats were still used at the start of navigation on the Yukon.

The western rivers (Mississippi) had boilers distinct from the rest of the marine industry. The first ones were just big externally fired cylinders. Flues were added to make them more efficient.

One type of return-flue boiler used in the Mississippi and on some of the first West Coast river boats consisted of two or more similar boiler drums connected in a common insulated setting. The fuel was burned in the furnace, which was external to the boiler drums. This furnace was only 4 or 5 ft. (1.2-1.5m.) long. The drums were usually 9 to 12 in. (22.8-30.5cm.) apart and formed the roof of the furnace. The hot gases from the brick furnace first passed over a bridgewall. This caused them to go along the bottom surface of the drums until the gases reached the combustion chamber, or back connection, which is common to all drums. From the combustion chamber the gases flowed in the opposite direction through the internal boiler

The fire door of the sternwheeler Moyie. She was a coal burner. The levers below the furnace controlled the grate and draft.

The smoke box on the after end of the boiler on the sternwheeler Moyie. This shows the tubes. The combustion gases and smoke from the fire on the forward end of the boiler went through the tubes and into the smoke box. Steam from the exhaust was combined with the gases on the way up the stack. This created a strong draft. This is a railroad type boiler.

flues to the common uptake. The uptake usually had a stack connected to each end of it. The gases then went up through the stacks. The uptake had a damper such as you might find on a wood burning kitchen stove to regulate the draft.

Each boiler drum usually contained one or more flues running lengthwise through the drum. The ends of these flues were secured to the ends of the drums. Both drums were connected to a common steam dome by steam nozzles. The steam dome provided a drier saturated steam as it was somewhat removed from the direct boiling action at the surface of the water. Steam is taken from the dome for the engine or other purposes. The water level was kept above the flues at all times. On western rivers' vessels a typical boiler was usually 20 to 30 ft. (6-9m.) long with 4 to 6 flues 10 to 14 in.(25.4-35.6cm.) in diameter. A large vessel could have a 40ft. (12m.) boiler. The diameter of the drums ranged from 24 to 30 in. (61-76cm.).

Suspended impurities and sediment in the feedwater (water put in boiler) settled to the bottom of the boiler drums and passed through the lower connections to the mud drum. The mud drum could be cleaned out with steam. A lot of these boilers had an outward look of a box-type boiler, but they were in fact a form of cylindrical boiler. They were generally fitted with a lever-type safety valve.

In an effort to get a lot of pressure at low cost these boilers were often poorly built. The low quality material and the poor workmanship caused a lot of problems and explosions.

Sternwheelers usually used locomotive boilers on the British Columbia rivers and lakes as well as the Columbia and other rivers to the south. Some of the later ones were designed for the river boat trade.

Locomotive type boiler on the sternwheeler W.T.Preston. Built in 1939. 149 tubes. Originally used bunker C but was converted to diesel. There are less BTU's in diesel but it is cleaner burning. The fire-room (stokehold) is unusual as the burners are on the aft end of the firebox and is located under the boiler.

Locomotive boilers are lighter than ordinary marine boilers. You can get high pressures from them but they are liable to prime (pick up water with the steam from the boiler) easily.

The locomotive boiler has a rectangular fire box enclosed in a shell of similar shape except the top is often semi-cylindrical. The furnace is surrounded by a narrow water filled space known as a water leg. A water bottom type had water under the furnace. The tubes are contained in a cylindrical barrel, extending horizontally from the firebox shell and ending at the smoke box.

Scotch Marine Boiler

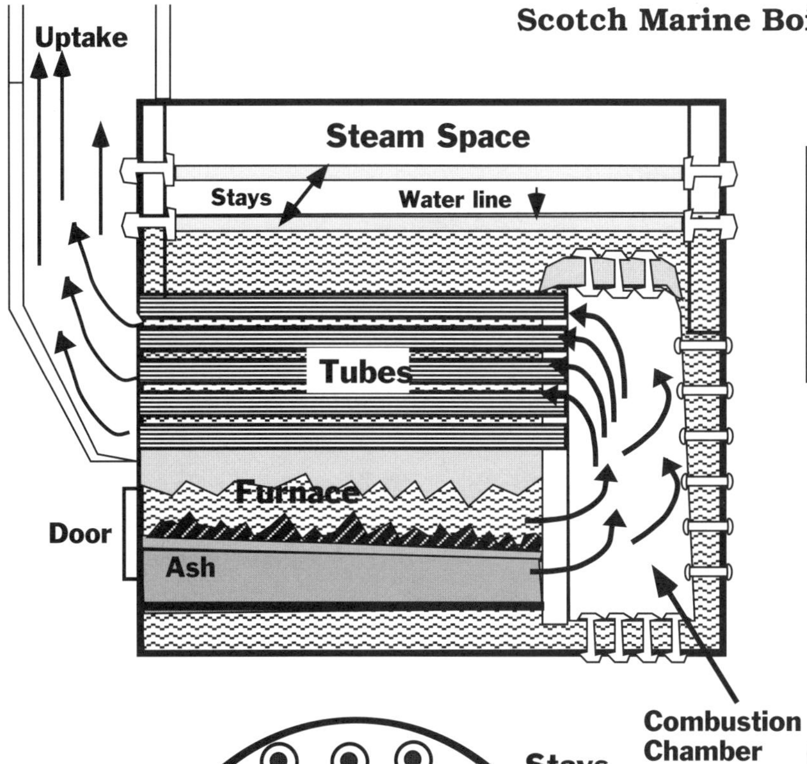

The scotch marine boiler. There were probably more of these boilers used on the Coast than any other type. It is a fire tube boiler. They were used by themselves on smaller vessels such as tugs or in banks on the larger ships. They were a relatively easy boiler to fire and were very reliable.

On newer scotch boilers each furnace and combustion chamber were separate.

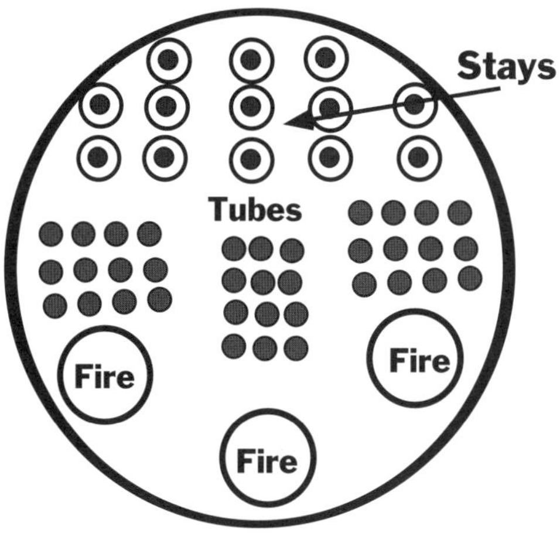

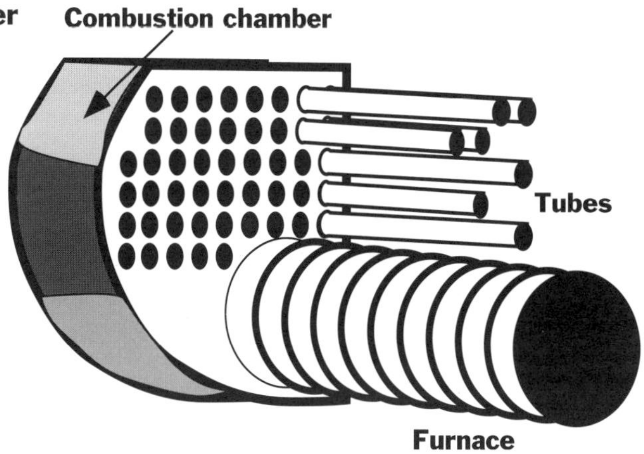

Scotch boilers were often double-ended for use on larger ships. Banks of these boilers could be fired from two or more stokeholds (fire rooms).

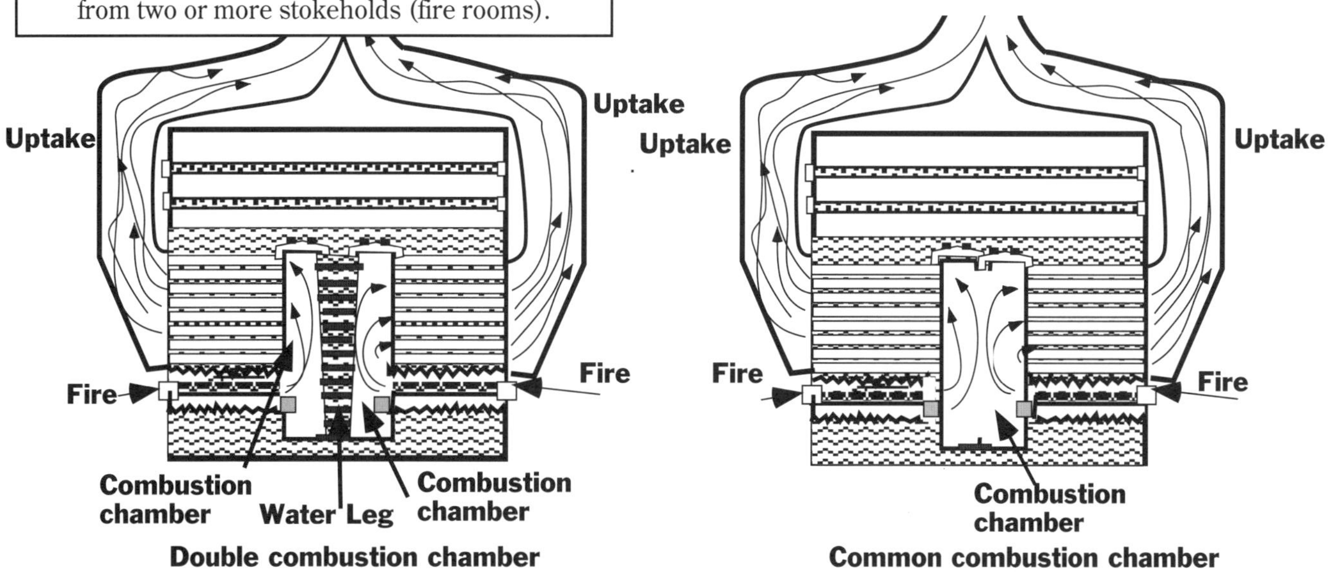

Double Ended Scotch

The tubes on the sidewheel ferry Eureka. The smoke box doors are open as they would be if the fireman was punching (cleaning) tubes. The tops of the furnaces are below.

The steam returning to the smoke box and exhausting up the long stack along with the smoke created a big draft. This is known as "blasting". On a locomotive the draft was so great that if the fireman threw sand in the firebox it would clean the tubes as it went through them.

They were fired with wood for years. When and if coal became available they switched to it. Some of the later ones were oil burners. Wood, a lot of it, was used on some rivers long after the coasters switched to other fuels. The Samson V's last boiler was a triple pass oil burner. Triple pass means the hot fumes went back and forth through the tubes three times.

Scotch Marine Boilers

The single-ended scotch boiler, or cylindrical boiler, was introduced in the middle of the 19th century. They became known as Scotch boilers because they were a very popular boiler in the Scotch shipyards. Reportedly the first one was built in 1862. Modern scotch boilers operate with a working pressure as high as 300 psi., but some years ago the working pressure was 60 psi. These boilers were originally coal fired but in later years were nearly all oil fired. There are wetback and dry-back models. You may find a steam chest mounted on top of one of these boilers. Larger ships had double-ended boilers that were fired from each end.

This is another type of internally fired tubular boiler. For some years most ships used this type of boiler. Gradually the faster ships switched to the more efficient water tube boiler. The size of a scotch boiler usually ranged from 10 to 18 ft. (3 to 5.5 m.) in diameter and 7 to 12 ft. (2.1 to 3.7 m.) long. The boiler had 2-4 furnaces. Large ships had several boilers and in some cases more than one stokehold. Single fired boilers were used on very small vessels.

The cylindrical drum or shell was made in one or more sections. The furnaces were corrugated and circular. A horse collar secured them to the rear tube sheet. Each furnace opened into a combustion chamber. The early scotch boilers opened into one common combustion chamber. On newer boilers the furnaces and combustion chambers were independent and surrounded by water. The components were held in place by stay bolts of various lengths.

The gases of combustion passed through the furnace to the combustion chamber where complete combustion took place. Most of the available heat was utilized before going through the tubes on its way to the stack. Without the combustion chamber a lot of the heat would have been lost up the stack.

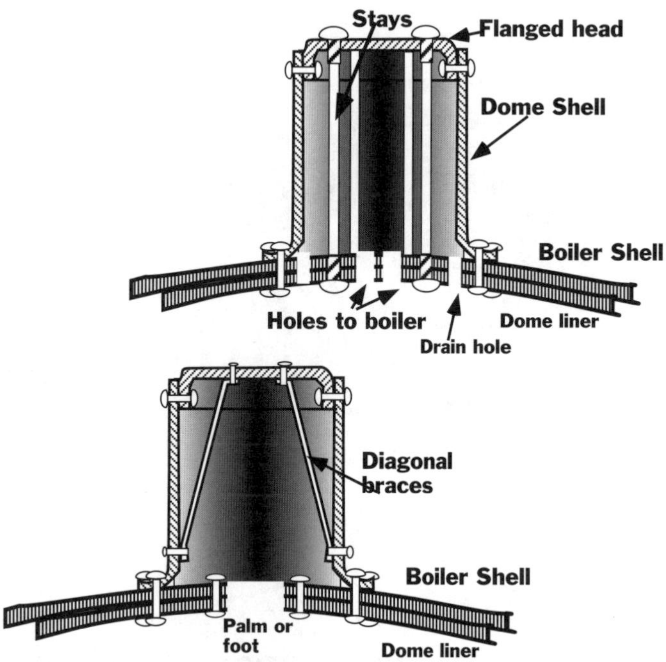

Various types of fire tube boilers had a steam dome to reduce priming. Priming occurs when water is picked up with the steam and delivered to the engine causing damage. The saturated steam is stored in the dome until it is used. The photograph is a steam dome off a shore based boiler but is similar to a marine one. Another type can be seen on the photograph of the W.T.Preston's boiler.

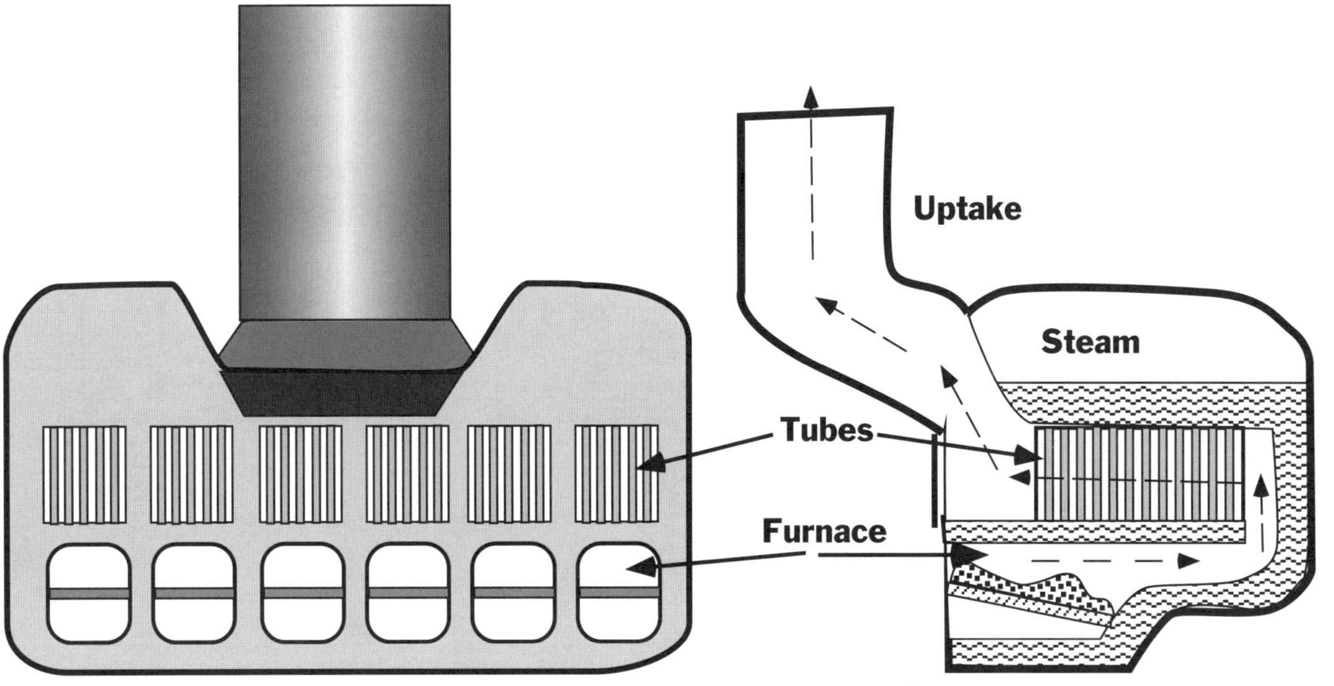

The vertical water tube boiler
One of the earlier types of water tube boilers. They were a low pressure boiler. They were popular with the U.S. Navy after 1857. Some of them were used on merchant passenger ships.
In order to be effective the water must rise in some tubes and must be returned in another set of tubes or headers for a good circulation. Some of the early water tube boilers did not do this so they were not too efficient.

On a number of boilers a fusible plug was fitted to the top of the combustion chamber Crown sheet that would blow if the temperature got too high due to low water.

The water directly above the furnaces became heated and rose among the tubes. The cooler water above descended to replace the rising hot water. The water at the bottom of the boiler was sluggish. This could cause stress on plates.

To improve the efficiency of the system superheated steam was used on later model boilers. To achieve this, saturated steam was heated again away from the main part of the boiler to produce a drier steam. This was very useful on turbine powered ships. You may find what seems to be extra tubes that were used for this purpose.

Water Tube

Water-tube boilers eventually replaced all the other types of boilers on faster ships. Small coasters and tugs were less likely to use them but there were some around. A water-tube boiler has a large number of tubes containing water. The tubes were heated by the furnace generating the steam. A lot of them had super-heaters and a number of drums. We can only describe briefly some of the types of these boilers.

They were less likely to explode and were fast steam generators. They were better for producing high pressure steam.

One of the early water-tube boilers was the vertical water-tube boiler. In the U.S.N. the principle steamers were fitted with these boilers after 1854. Some passenger vessels were fitted with them as well. It seems the U.S. had more success with early water-tube boilers than did England. They were more efficient when clean but they were hard to clean.

Most water-tube boilers now have one or more cylindrical drums on top, the highest being the steam drum. There are usually one or more similar drums at the bottom. The drums are connected with water filled tubes and sometimes hollow headers. Heat from the furnaces causes the water to circulate through the tubes and headers. The steam is collected in the steam drum.

Water-tube boilers are classified as large tube, small tube, horizontal, vertical and inclined. They have straight tubes, bent tubes, or a combination of the two. Large tube boilers often have 4in.(10.2 cm.) diameter tubes. Small tube, express boilers, have tubes around 1 1/2in. (3.8cm.) diameter. Express boilers operate under high pressure 350 to 1200 psi. By being able to operate at very high temperatures fuel consumption is reduced. The large tube boilers have less joints, are thicker, contain more water, and are not so likely to have all the water in them converted to steam. Small tube boilers generate steam more rapidly but are more difficult to clean.

Superheaters add heat to the steam after it has been separated from the boiling water. Superheated steam carries more heat energy per pound than saturated steam and reduces condensation of the steam in the turbine or engine.

As with boilers there have been many designs of superheaters. They are classed as either radiant or convection.

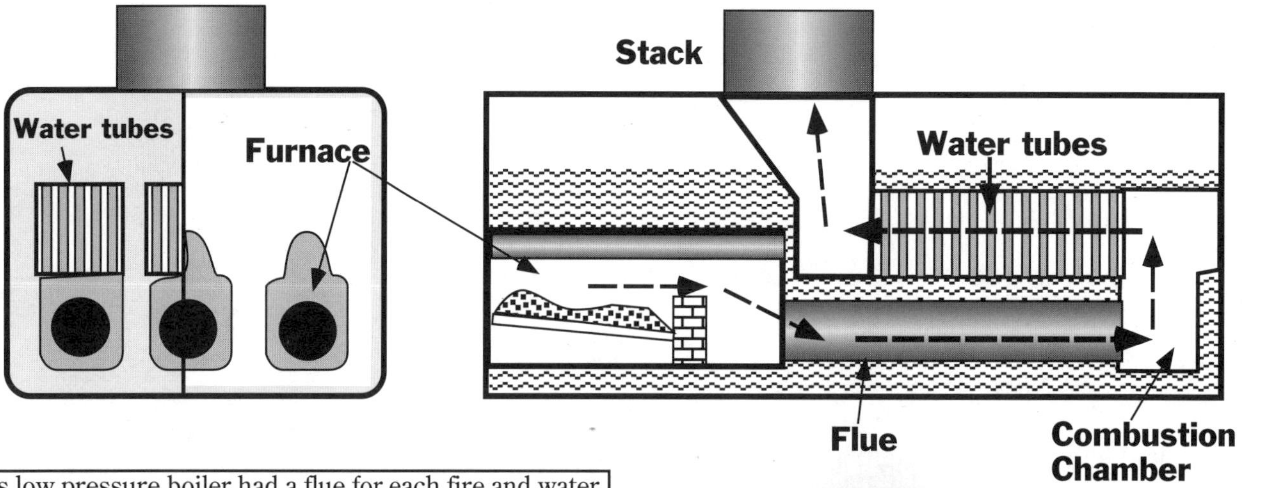

This low pressure boiler had a flue for each fire and water tubes.

Convection superheaters use the heat from furnace gases passing through them. They fall into two categories, interdeck or waste-heat. With the interdeck system the heater is in the boiler, while in the waste-heat system the heater is usually located in the stack.

Radiant heaters are located in the furnace. They can be wall, roof or floor type. They get their heat from the flame, hot bricks, or fuel bed.

Reheater superheaters reheat the superheated steam after it has been used. Another type is the separately heated superheater.

In the Babcock and Wilcox Boiler the generator tubes in which the water is heated connect to two rows of square headers. The headers and tubes are about 15 degrees from the vertical. Each front header is connected by a short tube to a large drum (steam drum). Each back header is connected to the steam drum by horizontal tubes (return tubes). Feed water is admitted to the steam drum. The for-

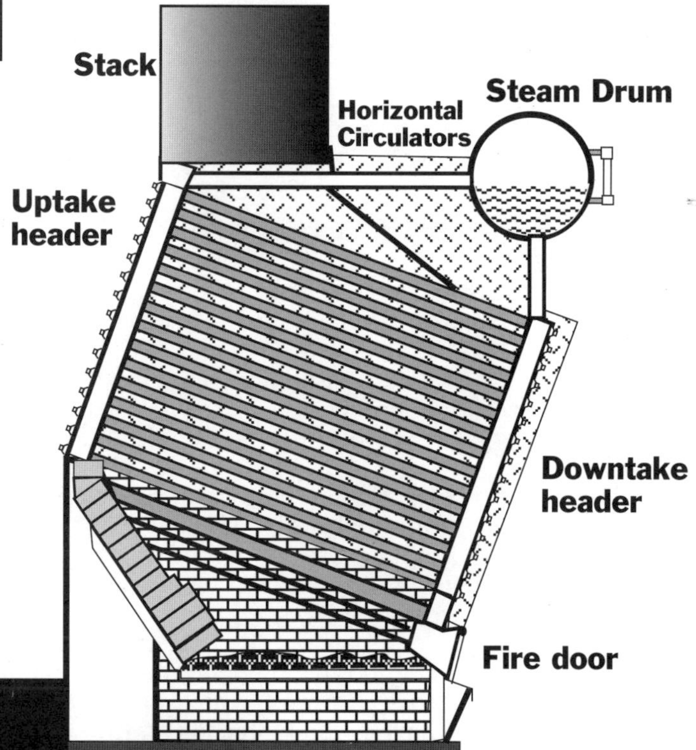

A early type of the Babcock and Wilcox boiler known as the B&W boiler. This was one of the earlier water tube boilers of the modern type. It was originally shore based but later became popular at sea after 1889. It was used on ships in one form or another for years. It has been used up to the present time.

The tube blowers can be seen on the side of the Portland's boiler. Steam could be blown on the outside of the tubes to clean soot and other material off them. The blowers could be rocked back and forth with the chains. As with all boilers keeping the heating surfaces clean improved the efficiency.

mation of steam in the generator tubes causes a circulation of water down the front headers, up the tubes and back headers and then to the steam drum.

Another popular type is the Yarrow water-tube boiler. This is a form of small tube boiler. It is useful for situations where quick steaming, light weight, high pressure and limited space are required. The boiler arrangement is triangular, in most cases the water tubes are straight except for the bottom row near the fire. The tubes connect circular water drums to steam drums. A lot of these boilers are equipped with superheaters.

The steam drum of a B&W boiler on the Portland as seen from the main deck. When she was operating she had two of these boilers. The large exhaust pipe can be seen entering the updraft. The main steam pipe is connected to the right side of the drum. Smaller pipes are going to assorted auxiliary equipment. Safety valves are located on the left side of the drum.
This type of boiler was used on many ships including Liberties.

The fire room (stokehold) is one deck down from the main deck on the sternwheel tug Portland. The boiler is oil fired.

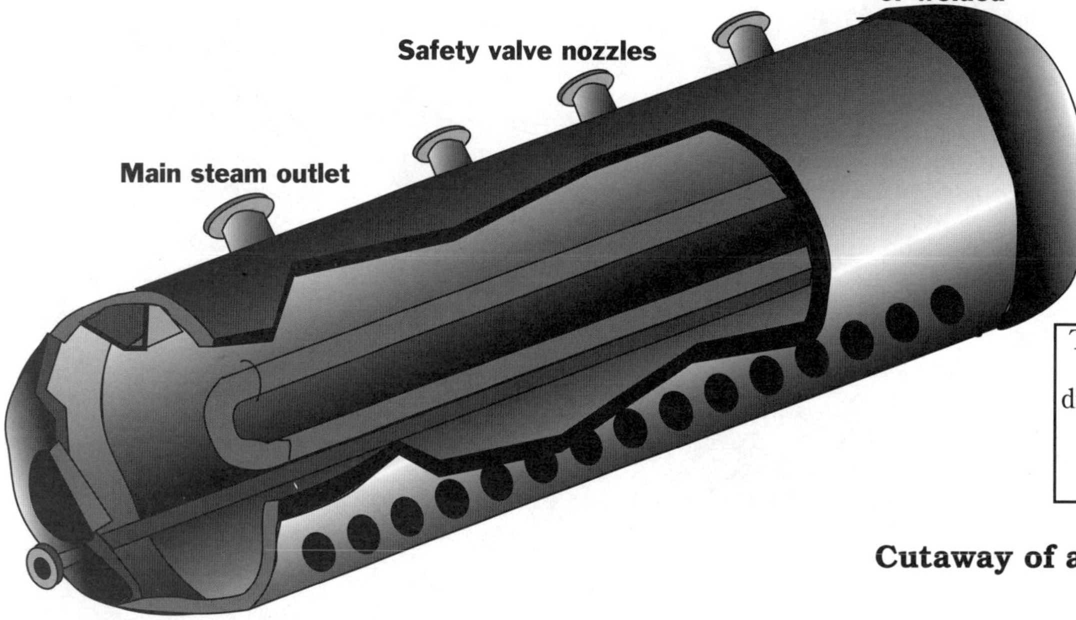

This is a cutaway of one type of steam drum. They varied in make and age of boiler.

Cutaway of a steam drum.

The yarrow type boiler.
The yarrow type boiler, sometimes called a three drum boiler, was very popular. There were many forms of this boiler produced. This is a simplified diagram of one of them. The principle was the same but the location of the parts and the shape varied a lot.
This type of boiler has an interdeck (inside boiler) convection superheater. Other types of water tube boilers included the B&J express and the Foster Wheeler boiler.

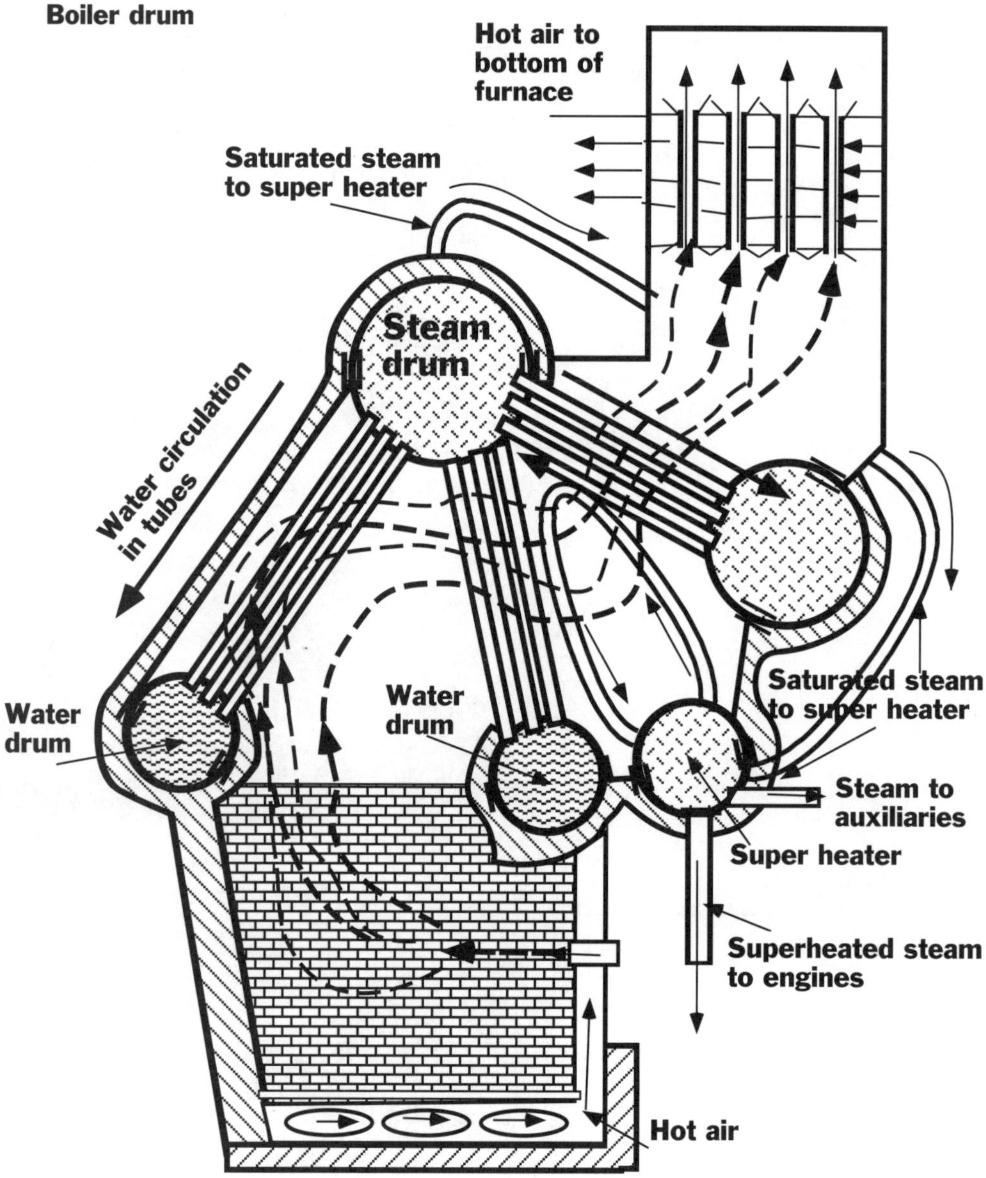

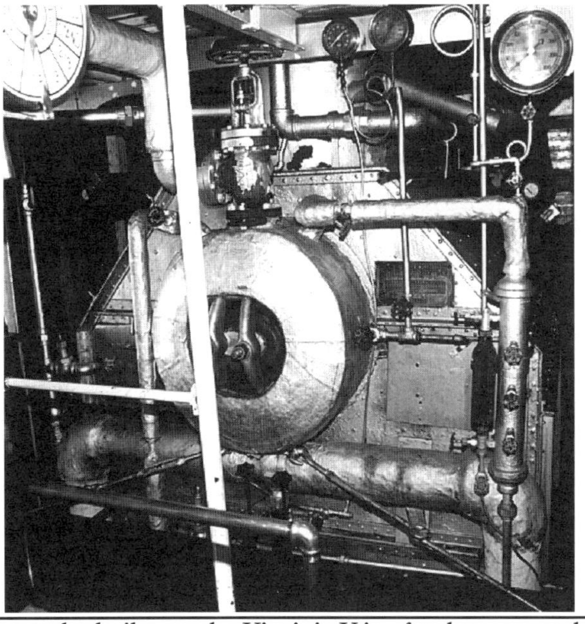

Watertube boiler on the Virginia V is of rather unusual construction. The steam drum can be seen in the center.

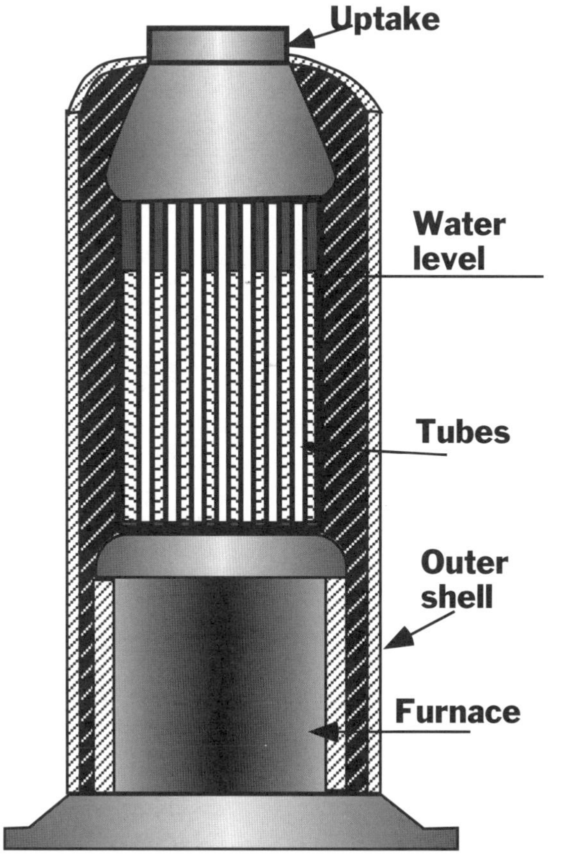

Advantages and comparison of water-tube and fire-tube boilers are: water-tube boilers are lighter, able to stand forcing, quickness to raise steam, less likely to explode and can operate with higher pressures; fire-tube boilers have less problems with feed water, are more rugged and it is easier to repair leaky tubes

Later steam ships used one form of oil firing or another. This required pumps to force the oil through the nozzles. The oil is atomized and injected into the firebox. As the bunker oil known as Bunker C is a heavy oil, it must be heated before it goes to the burners.

In addition to the more familiar natural draft boilers, forced draft stokeholds were introduced. This improved the efficiency of the boilers. This could be achieved by increasing the pressure in the stokehold or by the use of a fan connected to the furnace. Forced draft was introduced to the R.N. in 1882. This meant that engines could operate at higher speeds. There have been great improvements in boilers since WWII.

Every engine room and stokehold has a variety of pumps. These include feed-water pumps, saltwater circulating pumps, bilge pumps, ballast pumps, fresh water pumps and oil pumps. Originally these pumps were run by steam but later ships would have some electrical pumps as well. If the ship was a tanker or refrigeration ship (reefer) she would have numerous pumps aboard. This required masses of piping and valves. For this reason pipes and valves are often color coded. Valves are often marked.

A lot of the boilers and all steam pipes were covered with asbestos to prevent heat loss and reduce the danger of personnel injuries on the very hot pipes. This lagging was painted so very little asbestos got into the air except when repairs were made.

Vertical donkey boilers.
There were several types of them. This one had vertical tubes. Others had large horizontal water filled tubes or smaller tubes coiled around the combustion chamber (thimble boiler).Used for steam when the main boiler was shut down. Small boats such as launches used them for their main boiler. Sailing ships used them for heavy lift, reducing the size of the crew. They were used on docks with winches for moving cargo on sailing ships and early steamers.

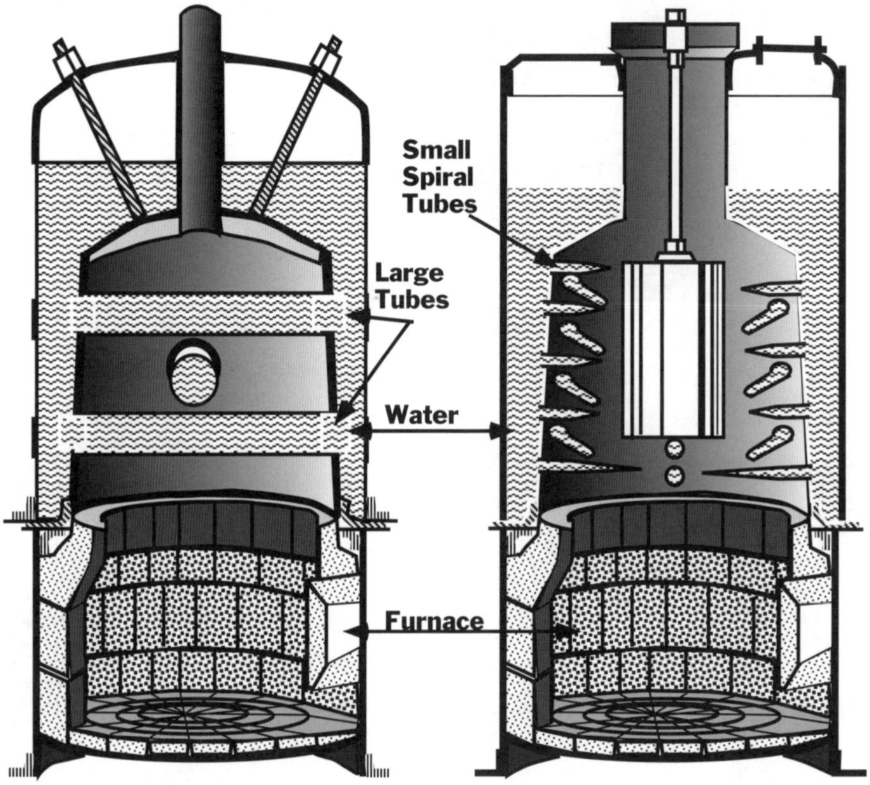

Vertical Donkey Boiler Cross Tube Type

Vertical Donkey Boiler Thimble Tube Type

Boiler Fittings

All boilers have a variety of valves and other fittings. The types and number of them depended on the make and age of the boiler.

Glass water gage - By looking at the level of water in the gage the fireman could tell the level of the water in the boiler. It needed blowing down to be accurate.

Test cocks - Gage cocks - On the front or shell of the boiler three small cocks were fitted. The distance between varied as to time and country of origin. One was located around 3 in. to 4 in. (7.6 to 10.2 cm.) above the heating surface. The second one at normal water level. The third one in the steam level. By opening either of these cocks the engineer could tell if the water was up to the level of the cock. It required some experience to make sure the reading was correct using either the cocks or the water gage.

Pressure Gage - Bordons type

Safety valve - Either spring or lever should let excess steam out of the boiler if the pressure got too high. The lever type was used at first and even if not tampered with could leak or stick. Superheaters should have their own safety valve.

Feed valves - Valves to admit water to the boiler. There were several types of automatic feed valves used as well.

Stop valves - These valves were located on top of the boiler to let steam go to the engine and the auxiliary equipment.

Scrum cock - Dirt floating on the surface of the boiler water that could be blown out.

Blow off valves

A hydrometer was used to check the amount of impurities in the water such as salt. Dirt and grease could be blown at the bottom of the boiler.

Separator - Steam from the main stop valve passed through the separator to remove any water from priming.

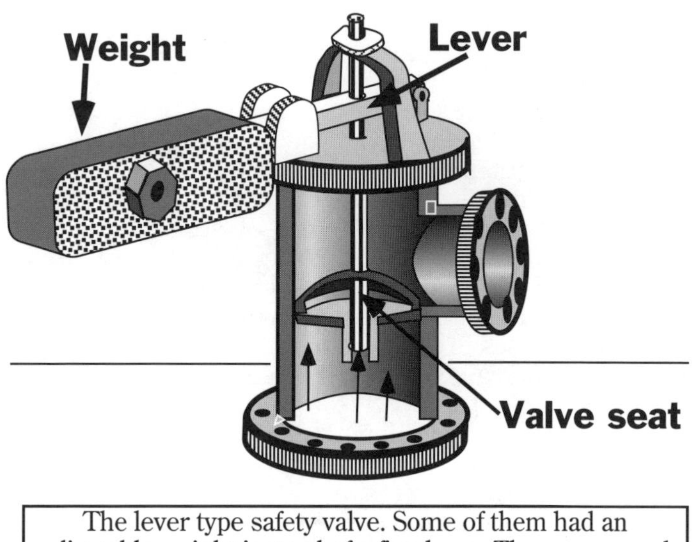

The lever type safety valve. Some of them had an adjustable weight instead of a fixed one. They were used for many years in one form or another.

The spring type safety valve.
Advantages it is not likely to stick, easy to overhaul,operated at any angle of heel. They totally replaced the lever type.

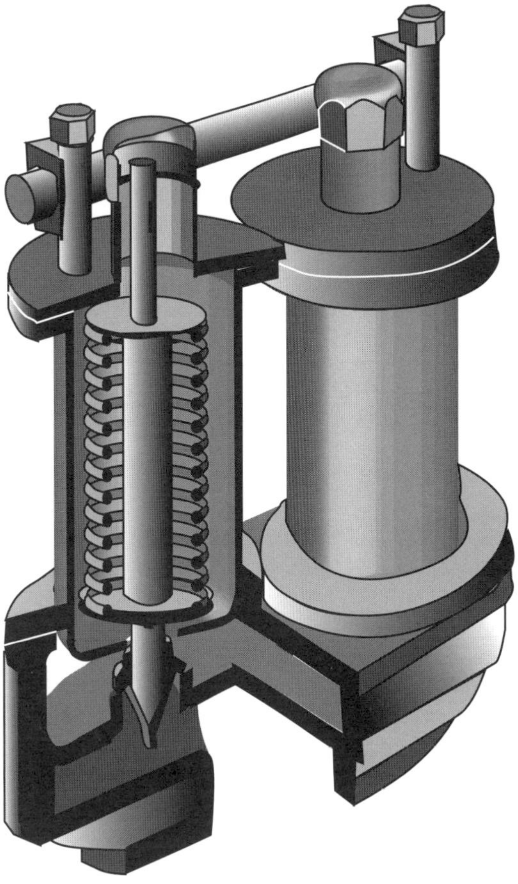

Boiler Safety Valves

Safety valves are usually required to be mounted in pairs.
A superheater should have its own safety.

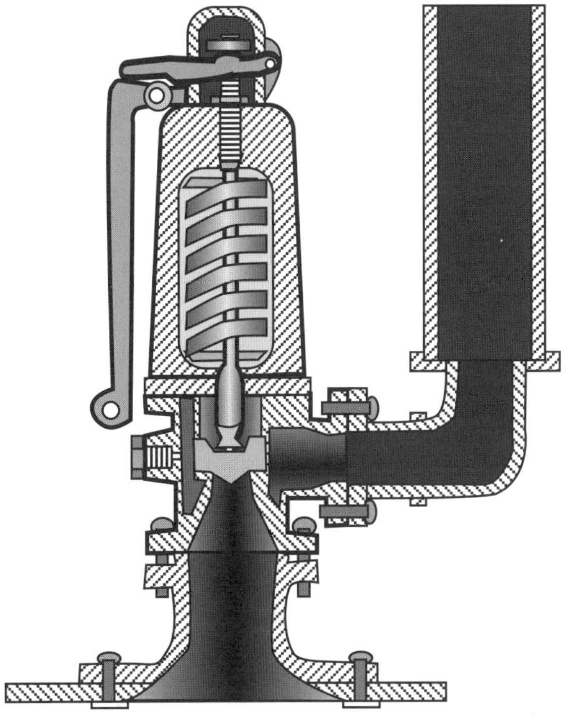

Another type of spring safety valve similar to the Crosby type. There are many regulations pertaining to safety valves. Each country has their own requirements for a particular type of boiler and when it was built. The safety valves on the right are on the W.T. Preston. Note the lever for blowing the valve.

Stoking

Regardless of the type of boiler used, firing efficiently requires a lot of skill and attention. In the days of wood and coal burning it was a lot of hard work. Wood was the only available fuel on the West Coast for some time until coal and later oil became available. At first coal was brought to San Francisco from England for the coastal trade. Wood was used in some remote areas even after coal was mined on the coast. The early low pressure boilers used an immense amount of fuel per day.

The efficiency (BTU's produced) of the type of available fuel was another factor. Depending on the type and quality of the fuel, approximately 1 lb. of coal produced 2 to 2.5 times as much heat as 1 lb. of wood. You would need 3 lbs. of coal to produce as much heat as 2 lbs. of oil.

These boilers required a larger number of coal passers and firemen. The crews on these first steam ships had to contend with working on a rolling ship with little light, heat and choking coal dust. Working conditions were never good on coal burners but they became a lot better with the advent of modern boilers and electric light.

Wood burning vessels required a lot of people on shore just to cut, fall and split the trees. The Beaver and some other vessels carried their own cutters but most of it was done by crews ashore.

In the off chance that some of the readers have not had the opportunity to cut, split and pile cordwood before chain saws were in vogue a quick description might be in order. After the tree is felled and limbed, slash properly piled of course, it has to be hauled to somewhere near a road. It is then cut to the correct length, by hand with the crosscut saw. It then has to be split. If the wood has a nice straight grain it splits easily with a splitting axe. If the grain is twisted and crooked you have to slab it with hammer and metal wedges. The split wood has to be piled neatly to dry. Wood with pitch in it has to be separated as the pitch would build up creosote in the furnace when burned. Though there were reports that pitch was thrown in the fire at times in order to win a race with a rival.

This is one of the common methods of using forced draught. Large coastal vessels used forced draught but most of the smaller vessels did not.

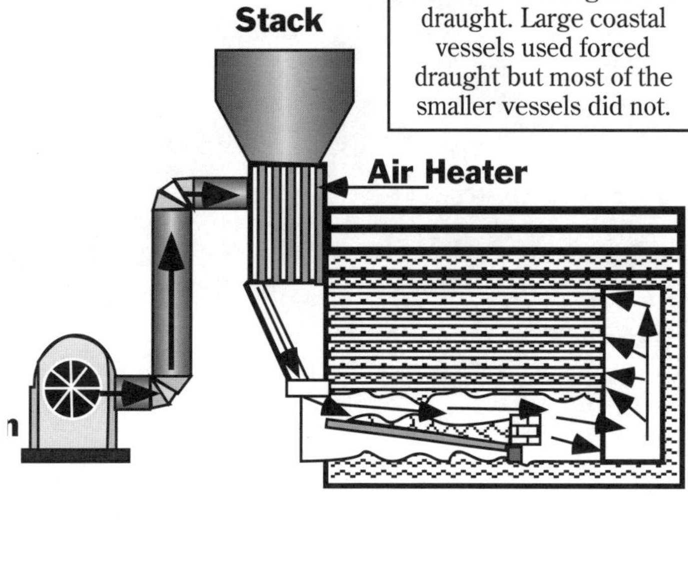

This wood had to be hauled to the landing or wharf to await the arrival of the steamer. On river boats and some other vessels, this wood had to be loaded on board by hand. A plank was run out to the landing and all the wood and cargo was packed up it. This was a very labor intensive operation. Wood was used up some of the rivers long after the coasters had switched to coal. In some cases the passengers had to help load the wood. A stern wheeler could easily use one to two cords an hour. A cord is a pile of stacked wood 4 ft. (1.2 m.) high, 4 ft. (1.2 m.) wide and 8 ft. (2.4 m.) long.

Except on river boats, coal burners in most cases could load coal using a chute from the dock. The coal still had to be stowed in the bunkers by hand. Every available space was used. In a lot of cases the vessel took on extra coal on deck. When at sea the coal had to be moved (trimming the bunkers) to the day bunker in the stokehold by coal-passers or trimmers. On some vessels the coal was brought to the stokehold on skids. Coal bunkers could arrive in the form of large lumps that had to be broken up with hammers before it could be used. In a rolling ship with hatches secured and the air thick with coal dust the men earned their pay. This is where the term "black gang" came from.

The forced draught blower on a Liberty ship.
The Jeremiah O'Brien - San Francisco Maritime Park

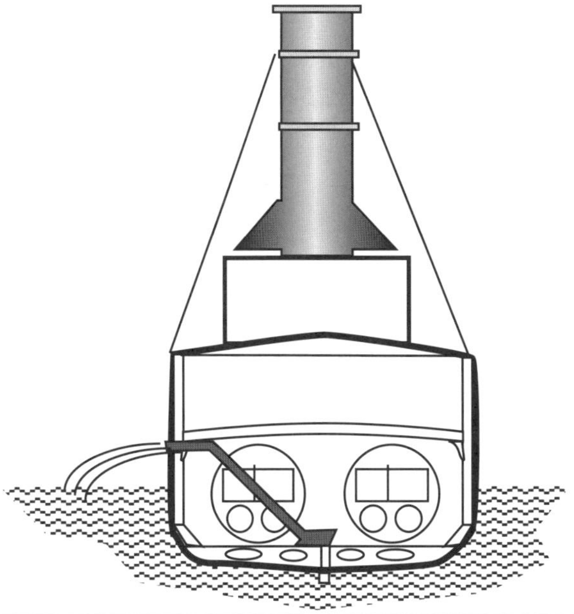

On a coal burner the ashes had to be dumped over the side. Larger ships had an ash ejector of one type or another. Small vessels used the bucket method. Regardless of the method used it was done on the lee side as the deck crew did not take kindly to their nice scrubbed decks and housework getting covered in ash.

The stokers or stoker, depending on the size of the ship, shoveled the coal from the day bunker to the fire. This was known as "making a pitch". On a scotch marine boiler the thickness of the coal was kept even, about four inches, over the entire grate. The grate was kept covered at all times. A rake was used to keep the burning coal at an even depth. Some engineers preferred to have more coal at the front of the fire. To reduce heat loss the furnace door was not opened for any longer than possible. With some coal, clinkers had to be broken up regularly with a slice bar.

Water tube boilers normally have brick lined furnaces. This one is in the Virginia V's boiler. Note the way the tubes run on the top of the furnace.

On forced draught ships the coal tends to burn more quickly at the sides and back of the furnace. The fuel must be pushed to the back before firing and possibly built up a bit more at the sides.

The fires had to be cleaned at regular intervals. The method differed to some degree. Basically the fire was somewhat burned down. The hot coals on one side of the fire were "winged over" to the other side with a slice bar. The clinker was broken up and pulled out with a rake. After some fresh coal was laid on the grate, the hot coals were winged over on top of them and the other side was cleaned in the same manner. On boilers with a number of fires the whole fire was pulled and started again with hot coals from one of the other fires. The ash from the ash pit was pulled as well. This had to be done without losing too much pressure on the boiler or the engineer had a few words to say to the fireman.

The hot ash and clinkers had to be cooled down with water after they were pulled from the fire. The ash was shoveled into an ash-ejector and forced over the side by water pressure. On small vessels the ash was hauled up in buckets and dumped over the side. Hopefully on the lee side.

Fires on a B&J boiler on a Liberty ship. These are forced draught boilers.

Objectives of firing

1 To avoid excessive fluctuations in steam pressure.

2 To obtain maximum heat from the fuel.

3 To avoid black smoke.

After gases are liberated from the burning coal, they must mix in the combustion chamber with the right amount of air at the right temperature if they are to be completely burned. The type of coal, how much it had weathered and the amount of impurities in it, had a lot to do with how well it fired and how much energy could be obtained per ton. In order to improve efficiency of the fuel, forced draft was used on some vessels even on early ships. Forced draft was not usually used on small coasters, river boats and tugs.

Coal consumption varied from ship to ship. An example of how much coal had to be shoveled, a triple expansion engine with a low pressure cylin-

The fires on the tug Master. Scotch marine boiler. Hot oil is pumped to the burners under pressure. The oil is sprayed into the furnace for combustion.

The pumps and heaters on the Master for heating the heavy fuel oil (Bunker C) and suppling it to the burner nozzles.

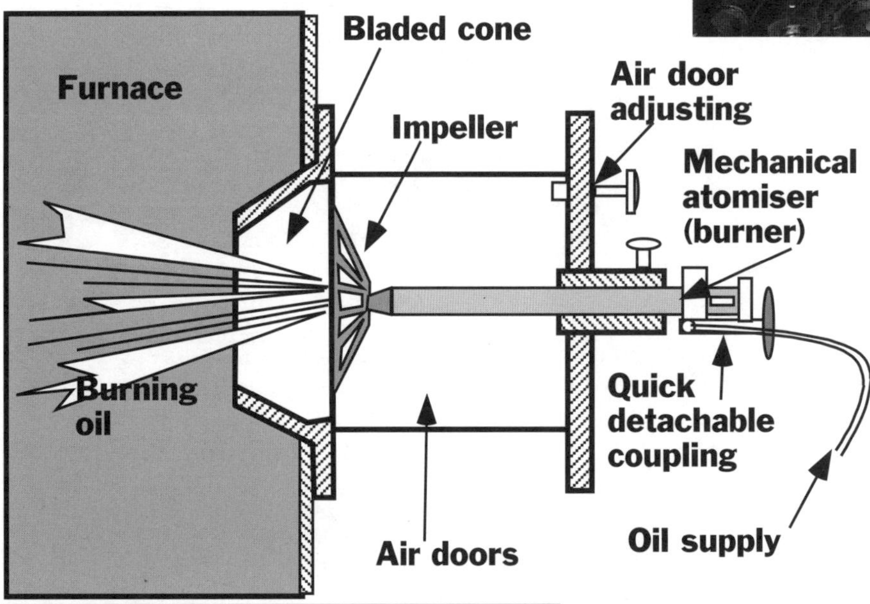

The types of burners and furnace doors changed over the years but this was a fairly common one.
Requirements for burning oil
1 Atomization must be thorough.
2 When atomized it must be brought in contact with the right amount of air for combustion.
3 The mixture must be burned in a furnace where the heat is radiated to help combustion.
4 The combustion must be completed before the gases come in contact with the heating surfaces.

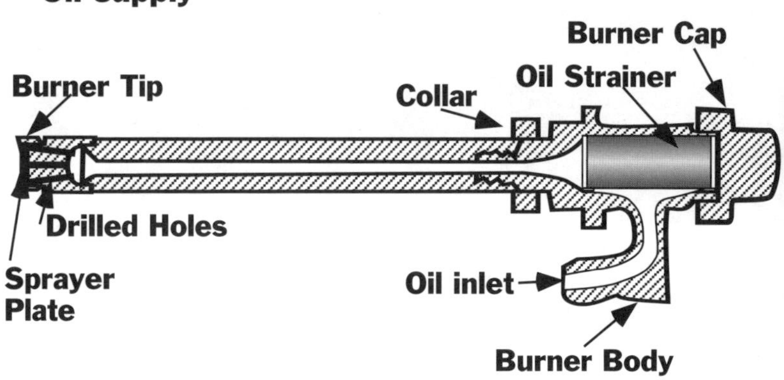

The fire on the sidewheel ferry Eureka. Scotch marine boilers with older type of burner, probably "peabody" type. Boiler operates at 30psi.

der of 60 inches diameter would use approximately 16 tons a day. By contrast a pair of side-lever engines, 80 inch cylinders and a 9 foot stroke used 48 tons a day. Some paddlewheelers used even more.

An oil burner still needed constant attention from the fireman. The heavy oil (flash point of over 150° F) had to be heated to between 100° F and 200° F before entering the sprayers. Air and the hot oil were sprayed into the furnace for combustion. Keeping the correct mixture of oil and oxygen demanded skill. The steam pressure in the boiler was controlled by putting on or shutting off sprayers. Each sprayer had its own valve for this purpose. In later times diesel oil could be used. This required less heating of the fuel before it went to the burners. This worked well for vessels on short runs as steam pressure for the heaters did not have to be kept up at all times. However the heavy oil was more efficient when the vessel was underway.

The oil-fuel system consisted of a suction valve on the fuel tank, suction stainer, oil-fuel pump, filters, heater, emergency valve, thermometer, sprayer control valves, sprayers, and combustion tube. All this equipment needed cleaning and general maintenance.

In addition to firing, the fireman had to keep an eye on the level of the water in the water gage. This gage could give a false reading and had to be blown down at times. Some boilers had feed water injected automatically but the boiler still had to be monitored.

On a fire tube boiler the tubes had to be blown and punched to get the soot off the walls of the tubes. The soot acted as an insulator and if not removed would make it difficult to keep steam up. Dirty tubes on either a fire or water tube boiler used more fuel. To blow tubes the door covering a set of tubes was opened. All the soot collected in the smoke box was blown up the stack. Then a long pipe was put down each tube blowing it out with steam. To punch tubes a wire brush was run down each tube. The outside of the tubes in water tube boilers had to be cleaned as well. This was often done with blowers. These blowers were located in various parts of the boiler. On a B & J boiler moveable tubes ran across the boiler at various heights. The ends protruded out one side of the boiler. The fireman could apply steam to these tubes and blow the soot and other debris off the water tubes. As they were limited in the arch they could turn, the steam would only hit the tubes and not the furnace area.

Needless to say, this could make an awful mess on deck if there was no wind to blow the soot away. The engineer was supposed to let the wheelhouse know when he was going to blow tubes so the vessel could be manoeuvered with the wind on the beam. This did not always happen causing not a little bad feeling from the seamen who cleaned up the mess.

The condition of the water in the boiler had to be monitored by the engineer. In the old jet condenser vessels salt water, mud and other impurities were major concerns. Even when modern surface condensers are used a leak can introduce salt water to the boiler. On some of the older boilers lime was added to the boiler and later sodium carbonate and sodium phosphate were used to reduce corrosion. Toward the end of the era the boiler water on ships was chemically treated.

The Engineer

The engineer on a steamer was not idle. On large ships there would be firemen and oilers to lubricate the machinery and keep everything clean. On smaller tugs and coasters the engineer would have to do a lot of the work himself, including firing in a lot of cases. Everything had to be monitored. The bearings had to be felt by hand to see if they were getting hot. Hot bearings were very common on the early engines. The rods had to be swabbed with lubricant. Some bearings had to be oiled by hand. Oil had to be kept in oil boxes or if the ship was fitted with lubricators making sure they were working. Working parts could get slack and need tightening. In addition, the engine pumps had to be run for bilges and the ship's water supply. Other gear such as steering equipment needed his attention. These were very practical men. They had to make repairs at sea often with very few tools. Machine shops were found on some larger vessels.

When the vessel was tied up in port the engine room crews often had a lot of work to do. Bearings had to be adjusted and leaky stuffing boxes attended to. At times the engine room crew would have to make major repairs. General maintenance required the paint brush and brass polish used nearly every day.

Cleaning a Scotch Marine boiler

The boilers had to be cleaned at least once a year. Scale consisting of carbonate of lime, sulphate of lime, magnesia and traces of salt had to be removed. The boiler was either blown down or cooled and pumped out. Letting the boiler cool down was considered the best method but it took time. Blowing down had to be done carefully as you could create strains on the boiler parts. The procedure varied on different ships. The boiler pressure was let down to about 50 psi. The ship's side cock was opened first then the blowdown cock on the boiler. After blowing down, the vacuum created had to be slowly relieved by opening an air cock before the manholes were knocked in. The top one was opened first to prevent any scalding water or gases from escaping.

Before anyone could go into the boiler to clean it out, it had to be ventilated thoroughly as there could be explosive gases inside. The boiler should have external heat applied inside to prevent condensation while the metal was cooled down.

When the boiler was cooled down and ventilated the fittings were removed and cleaned. Some lucky fellow got the job of going inside the boiler to clean the scale and any other foreign material out. This man, usually picked for his small size and lack of available funds, entered the boiler from the top. Depending on the size of the boiler he started at the top using wire brushes and worked his way down through the stays till he got to the top of the tubes. To get into areas he could not squeeze into, longer brushes were used. As one can imagine the air was not too wonderful and deep breathing was not recommended.

When he had all the scale in the top part of the boiler brushed down, he then went into the lower manhole and dragged out all the scale that dropped from above. He then reached up with his brushes and worked loose any scale he saw or felt.

When all this was done he started on the furnaces and combustion chamber. This was a dirty job to put it mildly.

After the engineer checked the job the boiler inspector was called down. If the cleaning job was rejected the engineer would be furious with the cleaner and probably suggested the office reduce his pay!

When all the fittings passed inspection and were reinstalled the boiler was filled with water. A small fire was started to slowly start warming up the boiler again. Depending on the pressure the boiler operated at and the amount of rush, it could take eighteen hours to get it up to steaming pressure again.

Sidewheelers

Low Pressure to Compound
Chapter 4

The First Steam Engines on the West Coast of North America

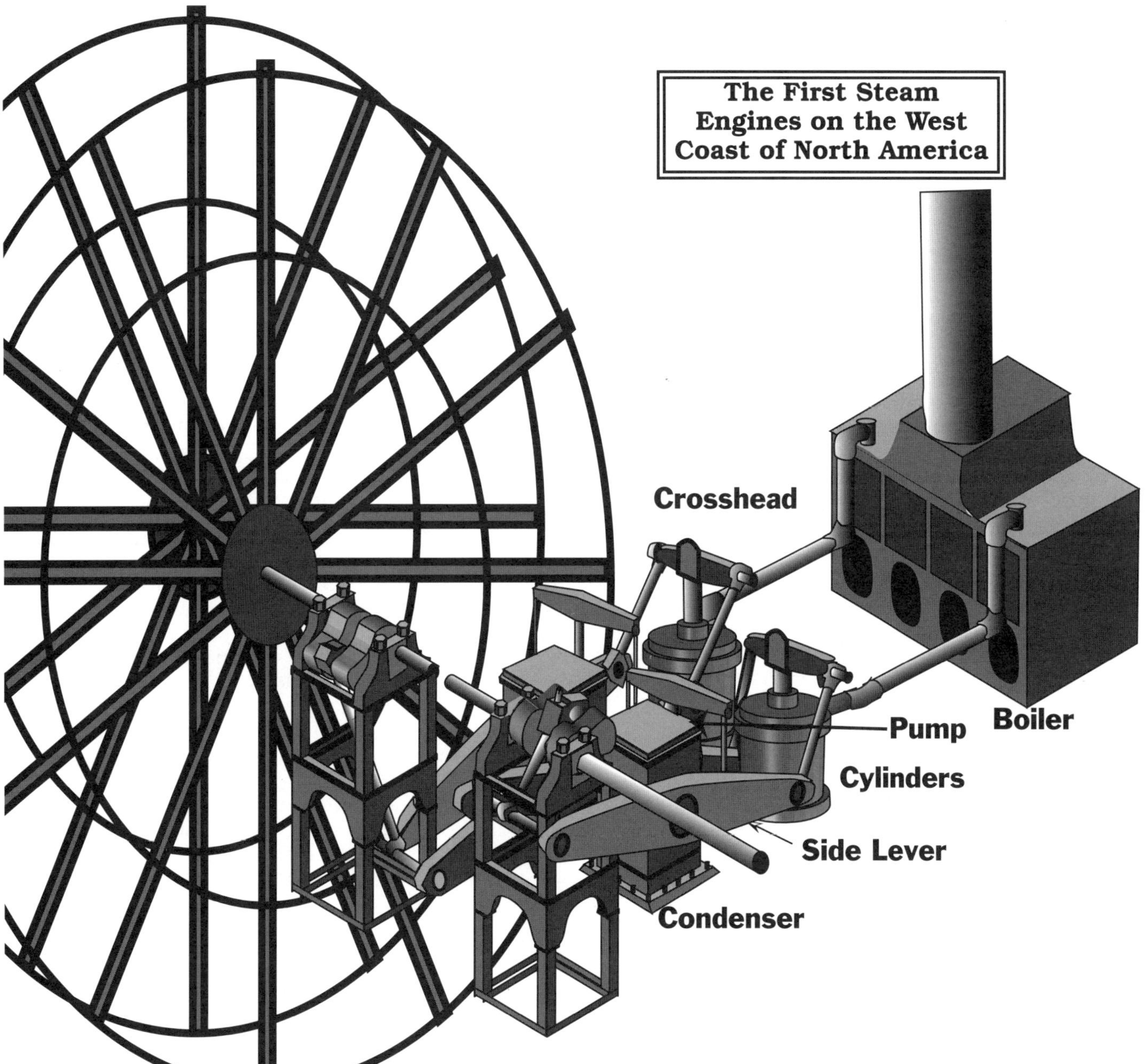

Engine Room layout of smaller ship equipped with side-lever engines. Box type low pressure boiler

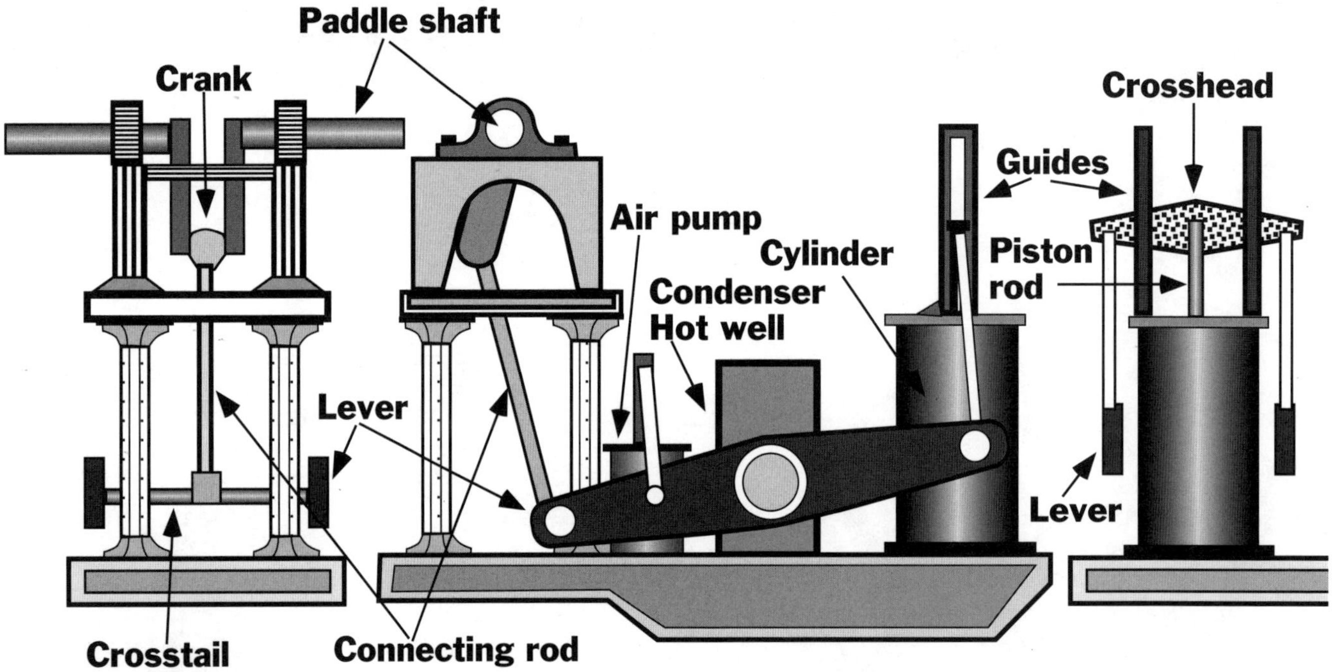

Diagram of a side-lever engine. There were many versions of this engine but the basic idea was the same. For reasons of clarity we have not shown the eccentric or cut off valves.The piston moved the levers on each side of the engine up and down. The levers were connected together at the opposite end by a crosstail. A connecting rod from the crosstail turned the crank.
The steam went from the cylinder to the jet condenser. The air pump created a vacuum and removed the water and air from the condenser and placed the water in the hot well until it was returned to the boiler.
For more on valves see Chapter 2 and page a6

Low Pressure Engines

The first steam engines were all low pressure engines and were used mainly on paddle-wheelers. Engineers were limited as metals and the methods of production could not produce a boiler that could accommodate much pressure. Crosshead engines were one of the first marine engines. They were somewhat patterned after English mine engines. At this time in the development of the steam engine, they were too big and slow to operate a screw (propeller) driven vessel.

The piston on these engines had to be directly under the crosshead, between the port and starboard sides of the frame. To reduce heat loss due to radiation, engineers discovered a longer stoke and larger cylinder sizes were more efficient. As time went on these engines became very large. The only way the large piston could be moved away from the frame was to use a lever to transmit power from the piston to the crank. Engineers had years of experience with the beam engines on shore. This knowledge was put to work on water.

Several types of engines using the lever were developed on both sides of the Atlantic. The side-lever and walking beam engines were the main types of low pressure engines that found their way to the West Coast. There were others such as the grasshopper, a popular U.K. tug boat engine that was, as far as we know, not used on the coast. Another low pressure engine, though not a lever type, used on the coast was the steeple engine.

Both slide valves and poppet valves were used in them. The U.S. used the poppet valve extensively. To increase efficiency cut-off valves were used on some engines.

Paddlewheel engines frequently had problems with broken cranks. In the U.S. the cranks were made by heating many small bars and hammering them with a steam hammer to make a weld. These welds often had undetectable flaws.

The materials and technology available to early engine designers were crude by modern standards. In "A Catechism Of Steam Engineering" 1857 there are several descriptions of using hemp and junk rings for packing pistons. When metals improved piston rings were used.

Side-Lever Engines

The side-lever engine was widely used from 1820 to 1860. Its low center of gravity was far better than vertical beam engines for ships exposed to open water. They were used on both sides of the Atlantic and a number of them sailed on the West Coast. The first one built in the Eastern U.S. was in 1846. They had been built in Canada in the 1830's or before.

Developed from Watt's beam engine, they were a low pressure double-acting engine. The engines were relatively large for the horsepower. A 500 hp. engine weighed 100 tons and needed 20ft. (6.1m.) of space. The 199ft. (60.6m.) California had a 70in. (1.8m.) diameter cylinder with an 8ft. (2.4m.) stroke.

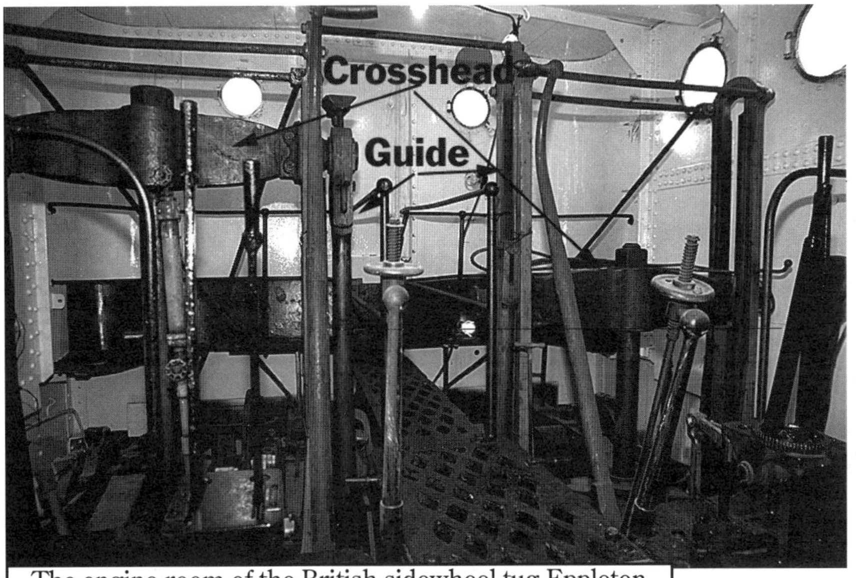

The engine room of the British sidewheel tug Eppleton Hall. A grasshopper engine.The guides and crossheads as well as the controls. She is now at the San Francisco Maritime historical park

Originally they operated at a very low pressure. Royal Navy vessels in the 1840's operated with 4 psi above atmosphere (14.75 + 4 = 18.75 absolute). The U.S. built engines originally operated with a steam pressure of 10-15 psi. The paddle wheels operated around 12 rpm. They used a lot of coal, 40 to 50 tons a day was not unusual.

Unlike the early vertical beam engines they were constructed entirely of iron. This made them a heavy but reliable engine. The main advantage, other than their low center of gravity, was a compact engine and all parts were below deck. This made lubrication and maintenance easier. A number of vessels had two engines situated side by side.

The cylinder, condenser, air pump and other parts were mounted on heavy bedplates. They had sockets for the iron pillars that supported the shaft. The bedplates often contained passages for steam and water. The castings were frequently ornate in the Gothic style. The condenser had pivots, called "the center," for the side-levers on them.

The steam was fed to the vertical cylinder from the boiler regulated by intake and exhaust valves. On English engines, slide valves controlled by loose eccentrics, were used while in the U.S. poppet valves were installed. When cut-off valves were used they were operated by a separate eccentric or cam.

Before the engine could be started a blow through valve had to be opened to let steam enter both sides of the piston condenser and air pump. The steam coming in contact with the cold metal condensed enough to form a vacuum. They were operating with such a low steam pressure that a vacuum was needed for the engine to start. When starting the engine the engineer operated the valves by hand with a starting lever until the engine was turning either ahead or astern. Then the grab (hook) was engaged to operate the valves automatically. The engine could only operate at a slow speed until the grab was engaged. On U.S. ships the order "hook on" was given when the ship was away.

A crosshead was fitted to the top of the piston rod controlled by guides. Connecting rods went to the two side-levers located on either side of the engine. The other end of the levers were connected by a bar called a crosstail. The crosstail had a connecting rod running up to the crank shaft. The paddle shaft (crank shaft) could be a split or single shaft. Early engine rooms did not have any auxiliary engines. The main engine operated all the pumps including the bilge pump if there was one.

The levers operated pumps with connecting rods and crossheads. The condensers were usually the jet type. Some surface condensers were used which reduced fuel consumption but the early ones had some problems. With the jet-type condenser cold water was mixed with the steam to cool it. This reduced the life of the boiler as salt water would be introduced in the boiler. The large air pump had a hot well on top of it. The air pump drew a mixture of water and vapor out of the condenser and sent it to the hot well. The feed water for the boiler was warm to reduce explosions. Molten tallow was used for lubrication on early engines which created problems for condensers and pumps.

The grasshopper engine. A form of side-lever engine. The levers swing from a fulcrum at one end of the lever rather than at the middle as with a true side-lever engine. A very reliable engine but not balanced enough for larger engines.

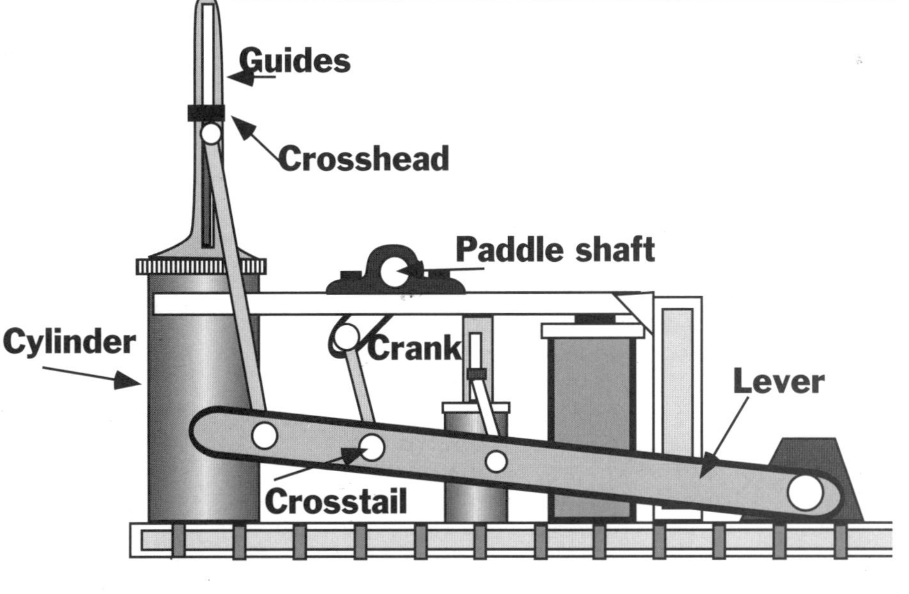

The side levers of the grasshopper engine. The rods coming down from the crossheads can be seen. There was not much room for the engineer to move.

A version of the side-lever engine was the grasshopper engine. The levers were of the third order. That is, the gudgeon or fulcrum was at one end of the lever instead of the center. These engines were not as balanced as the side-lever. When two of these engines were installed each engine could be operated separately for manoeuvering, or they could be hooked together when the vessel was running some distance. They took up less space than the side-lever. They were cheaper to build and permitted a longer stroke of piston in a shallow ship. They were very reliable and would keep working even when in poor condition. They were popular on English tugs and river steamers.

These engines though reliable were very heavy and fuel hungry. To give some indication of the advances made in steam engineering in the 19th century, the H.M.S. Salamander built in 1832 with side-lever engines weighing 275 tons developed 400 IHP. By 1900 a warship with the same weight of machinery would develop 3000 IHP with one quarter of the fuel consumption and utilizing a lot less machinery space.

The last British side-lever engines were built around 1860 for the La Plata, Arabia and Scotia. Each engines' cylinders were 100in. (2.5m.) in diameter with 12ft. (3.7m.) stroke developed approximately 4632 IHP. Steam exhausted to a jet condenser located between the cylinder and air pump. The last side-lever engines on a British ship were installed in 1861 in the Scotia.

The Vertical Beam Engine - "Walking Beam Engine"

The vertical beam engine or "walking beam" engine as it was known was a uniquely American marine engine. Patterned after English pumping engines, they originated in the rivers and bays of the East Coast. The walking beam engine started taking over from the side-lever engines on U.S. deep water vessels in 1851. There were many of these used in passenger vessels all over the West Coast. Some were on runs taking them across the Pacific. When looking at old pictures of sidewheelers operating on both sides of the border, the tell tale beam can be seen above the housework. The walking beam engine was a simple reliable engine and cheap to build.

One advantage was they did not need as ridged a hull as with other engines. When in a sea the frame could move a bit transversely without affecting the engine. This worked well with the wooden hulls. They lasted for years but were hard on fuel.

They were a large single cylinder double-acting beam engine. A lot of the principles that applied to the side-lever applied to them.

The walking beam engine had a large diamond shaped beam mounted on top of a large A frame structure. While the beam was usually seen above the top deck, it could be enclosed by a cabin for protection. On some larger sidewheelers two engines would be installed side by side for greater power and manoeuverability.

It is generally thought that the first walking beam engine built in the U.S. was in 1820 though there is some evidence of earlier ones, the last was built in 1924. They were big engines. The average height was between 35ft. (10.7m.) to 45ft. (13.7m.) high. The larger ones measured 60ft. (18.3m.) from the beam to the bed plate. As you can imagine, they were very heavy as well. The beam alone frequently weighed upwards of 20 tons. Reportedly the largest one developed 7500 hp. with a 110in. (2.8m.) cylinder and a 14ft. (4.3m.) stroke.

There were many sizes of these engines built. In the 1860's the U.S.N. did tests on a number of steamers with this type of engine. One of them was

The port cylinder head of the Eppleton Hall. The crank (paddle) shaft and eccentric can be seen just ahead of the cylinder. This shows how compact these engines were.

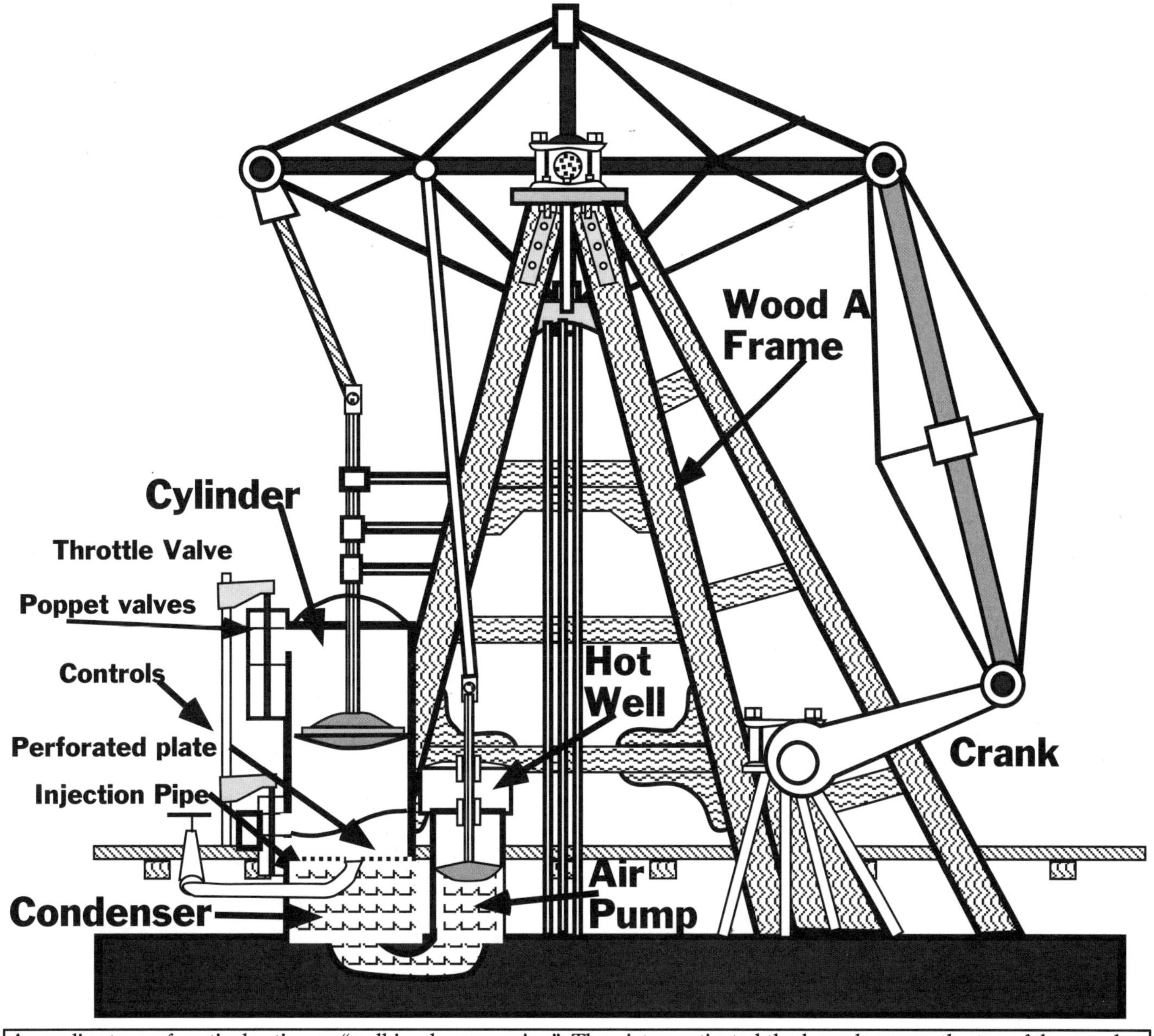

An earlier type of vertical action or "walking beam engine". The piston activated the large beam on the top of the wooden frame. A wooden connecting rod, strengthened with metal rods turned the crank. Poppet valves regulated the entry and the exhaust of the steam in the large cylinder. The steam was condensed with a jet condenser.

the 200 ft. (61m.) Georgeanna. She displaced 539.8 tons and supported 28.5 ft. (8.7m.) paddle wheels. She had fire tube boilers and operated around 30psi. The diameter of the cylinder was 44in.(1.12m.). Her piston stroke was 11ft.(3.4m.). She had poppet valves and cutoff valves. These tests were done on average coastal vessels. Boiler pressure was around 40psi. The pistons were between 42in. (1m.) to 58in. (1.5m.) with strokes of 10ft. (3m.) to 12ft. (3.7m.) turning 21ft. (6.4m.) to 28ft. (8.5m.) paddles. On the West Coast the engines varied from the small coaster with a cylinder of 30in. (.76m.) to large passenger vessels on the Panama run with cylinders of 105in. (2.67m.) and stroke 12ft. (3.7m.).

The early ones operated with only a few pounds of steam pressure. By the end of their use they would work with 30 to 70 psi. It was still considered a low pressure engine.

At first the A frame, beam, and engine base were constructed of wood. Newer engines were constructed entirely of metal. The wooden frame and beams were reinforced with iron rods. Some of these rods ran from the side of the frame down to the bottom of the hull. Some engines like the one on the ferry Eureka had her wooden frame replaced with a steel one.

Steam from the boilers was stored in a steam chest connected to the cylinder. The steam was introduced to the cylinder by a series of valves. The piston rod was connected to one end of the beam by rods working off the crosshead. The beam was mounted on an A frame constructed of wood or iron. The up-and-down movement of the piston rod caused the beam to rock back and forth. The other end of the beam had a long connecting rod which turned the crank on the main paddlewheel shaft. In addition, rods from the beam operated pumps and other equipment.

The connecting rod and walking beam bearings were made of brass or bronze. There were two rods side by side going to a crosshead on the piston side. The four shaft support bearings, two inboard on the engine bed and two outboard on the hulls' sheer timbers were massive iron castings lined with brass or babbitt bearing metal. On some vessels the shaft bearings on the hull could be adjusted to keep the shaft aligned with the hull. Bearings on the outboard ends of the crosshead were flat blocks of brass or bronze.

The intake and exhaust valves were single "poppet" valves on the earlier engines. When steam pressure increased, it was necessary to use double-poppet valves. As with other steam engines there were a number of different valve systems used over the years. Eccentric bearings and rods were used to control these valves. This determined which way the engine ran. Depending on the load, adjustable cut-off valves were used to make the engines more efficient. Variations of the Sickels' and the Stevens' cut-off systems were used. Some vessels used double eccentrics to cut the steam off before it had fully expanded, but these were not adjustable and could create a hesitation in the speed of the paddle wheel.

Certain vessels had a loose eccentric reversing gear while others had separate eccentrics for ahead and astern. The loose eccentric consists of a single eccentric, which is free to turn on the shaft and a disk keyed to the shaft. The disk has a circular grove with a sliding block attached to the eccentric. The ends of this groove limit the position of the

The steam boxes, rocker arm, lifters, wipers and gages on the ferry Eureka. She is part of the San Francisco Maritime National Historical Park.

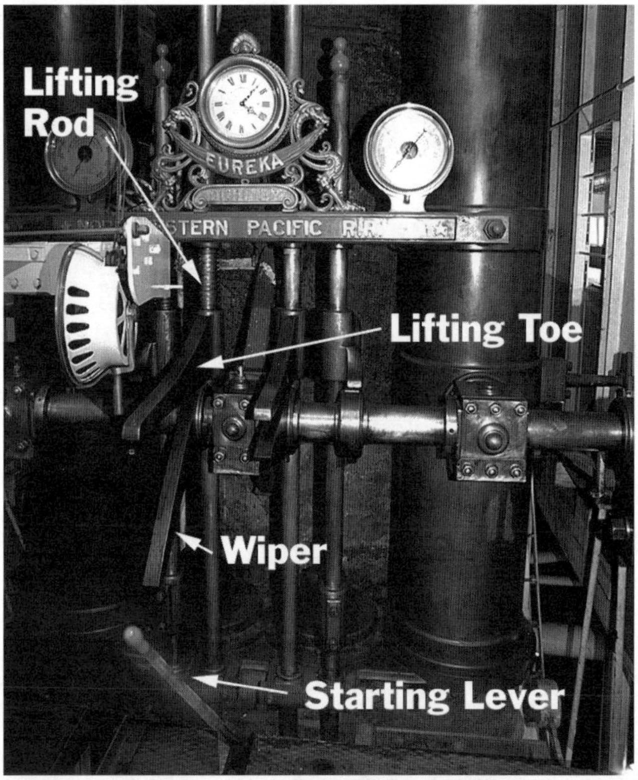

The steam (left side) and exhaust (right side) valves are operated by the engineer with the starting lever until the engine is going. The hook is engaged and the eccentrics operate the valves.

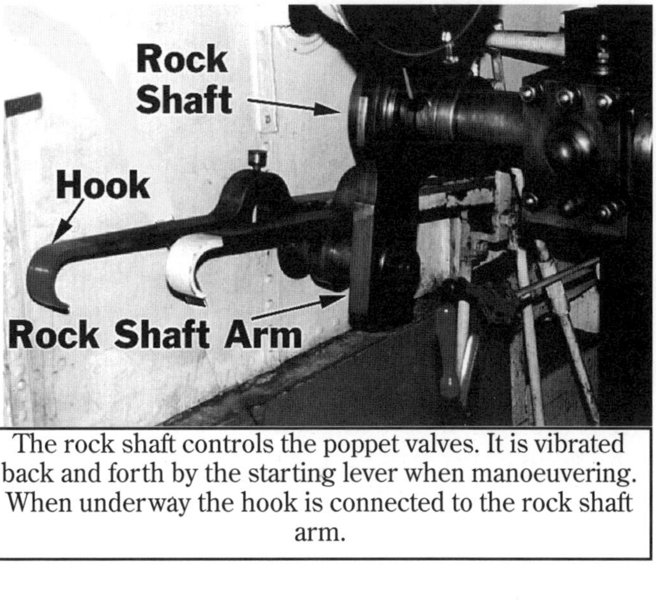

The rock shaft controls the poppet valves. It is vibrated back and forth by the starting lever when manoeuvering. When underway the hook is connected to the rock shaft arm.

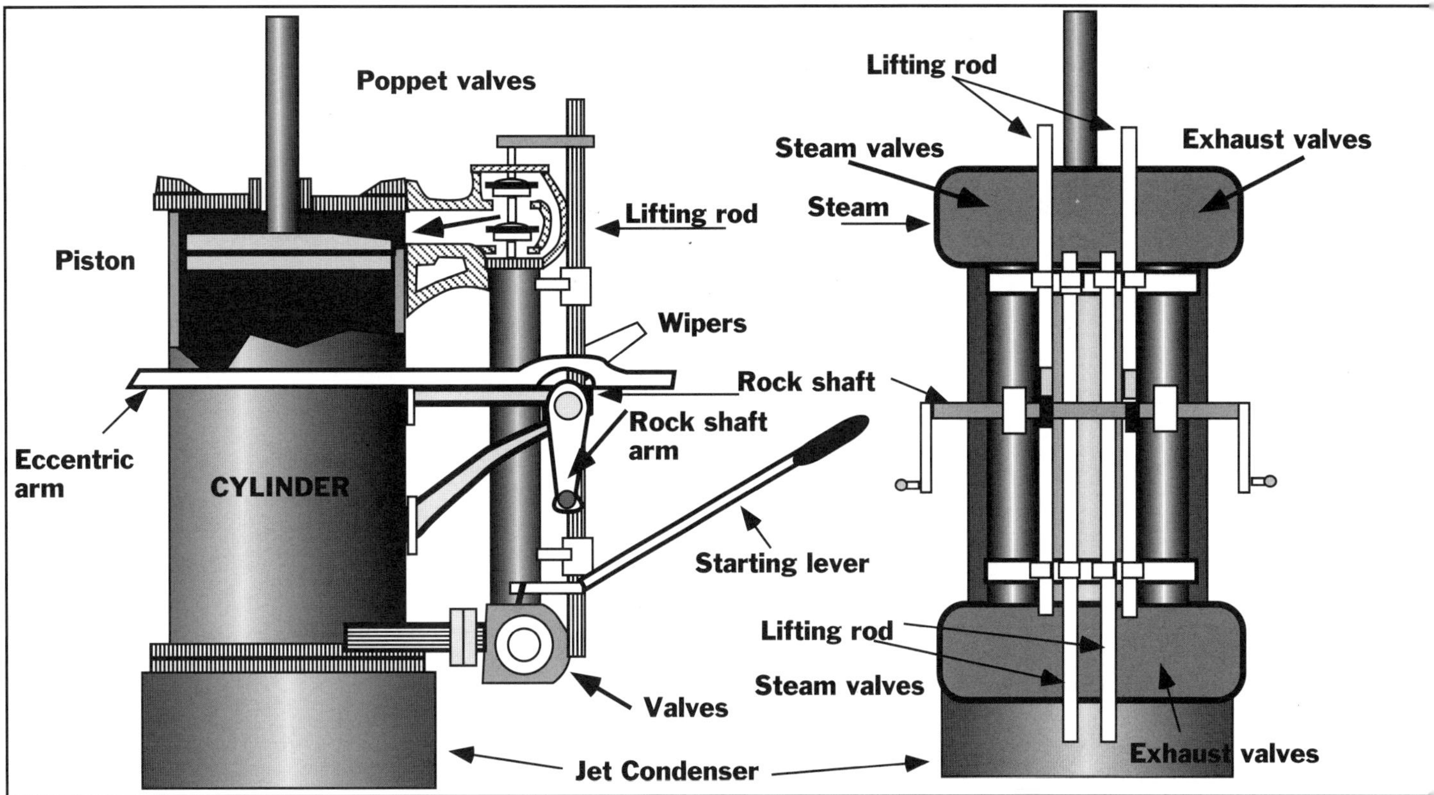

Valves and controls on a walking beam engine. For clarity some parts have not been included in this illustration.

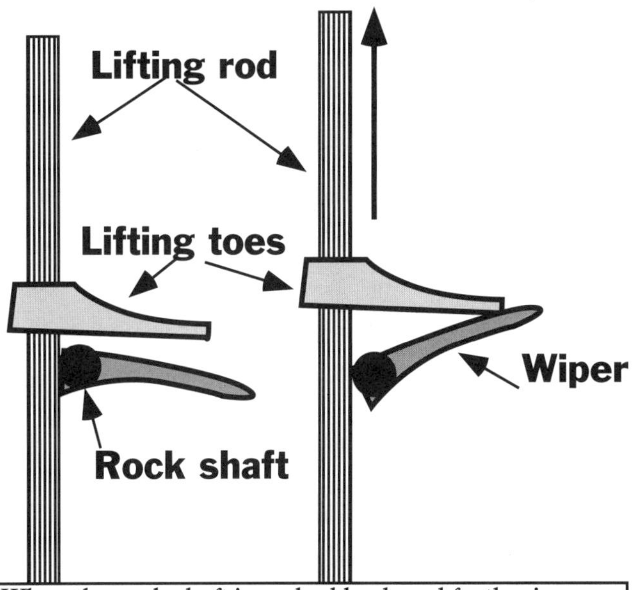

When the rock shaft is rocked back and forth wipers engage the lifting toes to operate the lifting rods. There are a number of these rods to operate the poppet valves on the top and bottom of the large cylinder.

eccentric to the proper angle it has to make with the crank.

Walking beam engines used jet condensers in most cases, where cold water was mixed with the steam. This created the same difficulty with salt water at sea and mud in the rivers entering the boiler. The steam being cooled created a vacuum. This gave more power to the engine due to a greater pressure difference between the exhaust and the steam entering the cylinder. The condenser also saved water.

In later years the surface condenser was introduced. It was a drum-like container with many tubes full of moving seawater. This cooled the steam without the seawater entering the system. This greatly improved the life of the boiler. They were eventually the only type of condenser used in salt water.

They were fitted with a single-acting air pump to draw out the water and steam. This had a slender rod connected to the beam. The air pump had a "hot well" mounted on top of it. Air consisted of a mixture of water and vapor originating in the condenser. As the seawater wound up in the boiler it was first warmed up in the hot well. The capacity of hot wells was from 100 to 200 gallons. A hot well of that size was not usually required with the surface condenser.

These engines had a low operating speed. Some engines operated around 17 to 25 rpm. With proper care they would have a life of over 30 years.

As with other steam engines, the engineer played a major part in the operation of the ship. There were no single-lever air controls here. You could not go flying into a dock and slam it full astern at the last moment for a spectacular high speed landing, not twice with the same dock anyway. They were very difficult engines to operate. Everything from the amount of steam to the amount of water injected into the condenser required much experience. You could get the engine stuck on top center. This stopped all manoeuvers

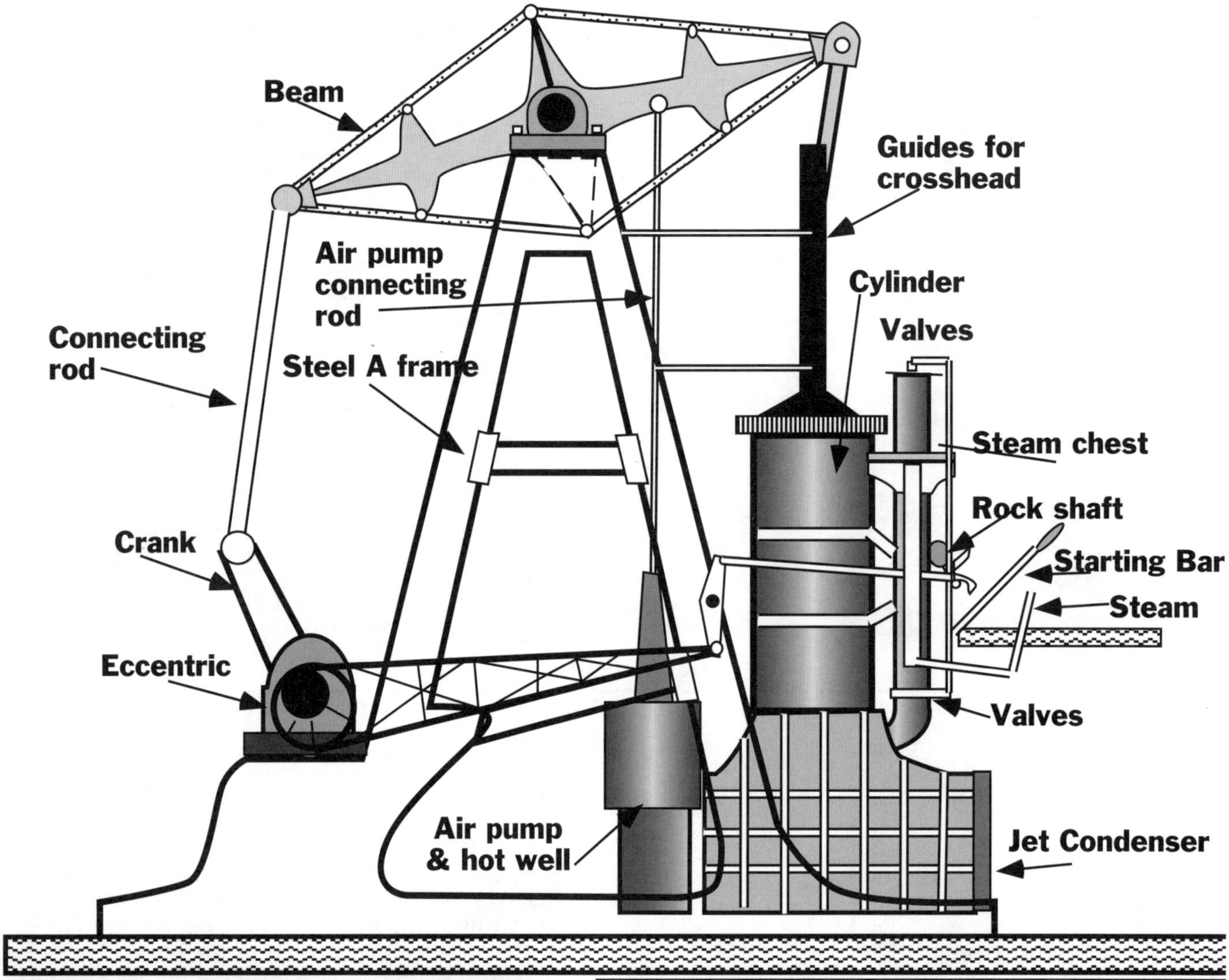

Steel A Frame - Walking Beam Engine

Above Beam of the Eureka
Left- Air pump and hot well. The bottom of the cylinder can be seen at the right.

Crosshead and guides for the main connecting rod and the short rods to the beam. Most engines just had a single connecting rod to the beam.

The large bearing the beam swung on.

till the crank shaft could be moved. There were a variety of ways to get the crank moved but they all took time. A long bar inserted in the paddles or spokes to turn the paddle wheel was the most usual way. If there was steam pressure remaining in the cylinder it could get rough on those doing the barring.

The engine was started by using a starting bar that operated the valves by hand. This was quite a common method for a lot of the older engines. The engineer had to know the position of the paddle shaft so he would know which valves to open first in order to get the paddle wheel to turn in the required direction. When the engine was running the eccentric rods were lowered on the hooks so the rocker arms would operate the valves automatically.

A number of these engines were built in San Francisco. They were used in the U.S. and Canada.

The bearings on the beam for the connecting rod

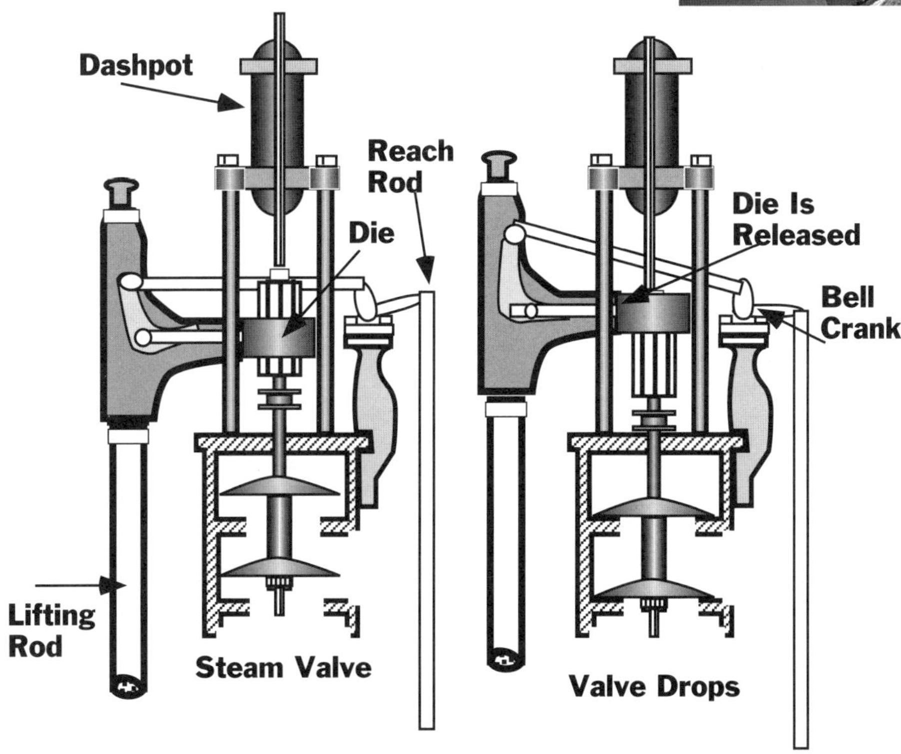

The Sickels' Cut-Off Valve
The Sickels' cut-off and the Stevens' cut-off valves were the most common systems used on Vertical Beam Engines. The Stevens' cut-off used a separate rock shaft and eccentric for the steam valves.
With the Sickels' valve the die lifted the valve stem. The adjustable bell crank released the valve in the die dropping the valve to cut off the steam. The dashpot was a type of buffer so the valve would not hit too hard. Both types of valves were adjustable to some extent.

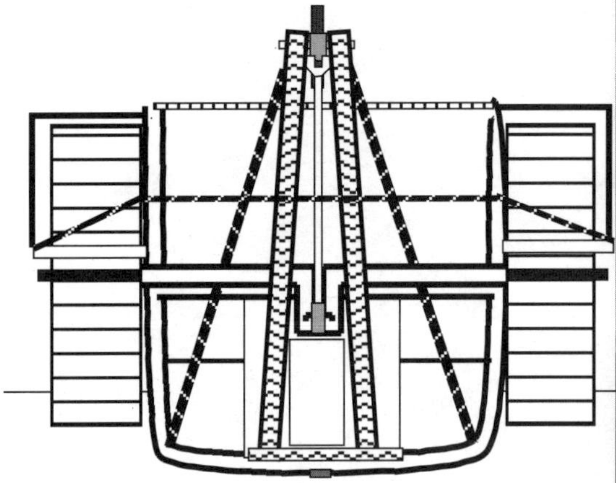

The walking beam engine had to be supported by rods to reduce thwartship's movement.

Above -The connecting rod to the air pump.
Right - The connecting rod between the crank (bottom) and the beam. Note how long the connecting rod is. The crank is at the bottom.

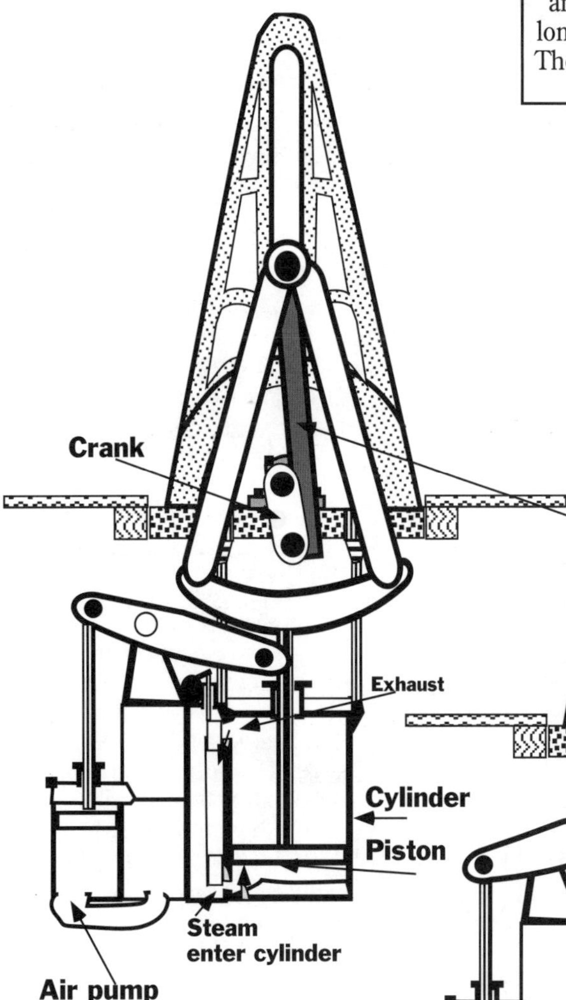

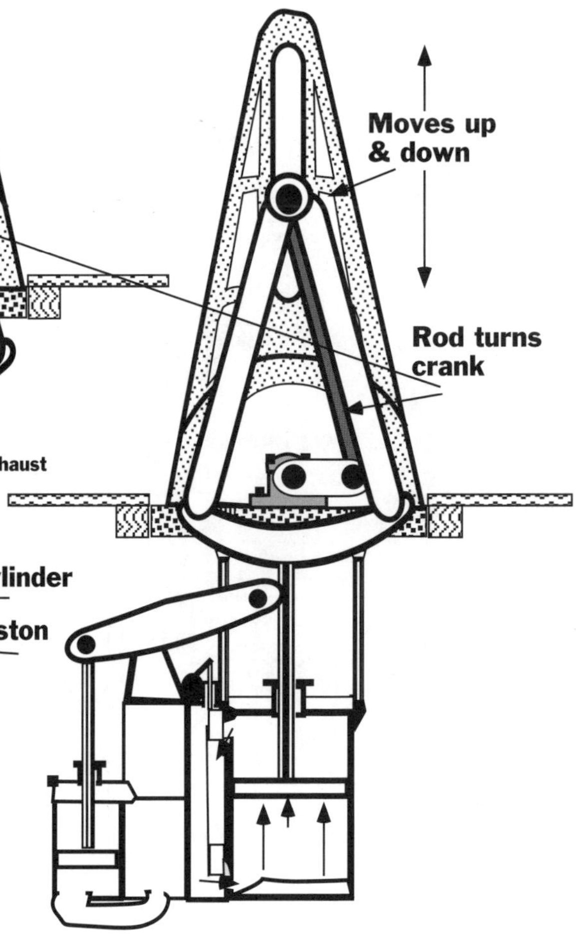

Steeple Engine

Steeple Engine-Low Pressure

Another type of low pressure engine was the steeple engine. It was one of the earliest engines to have the piston directly under the crank. Piston rods were connected to crossheads. The connecting rod ran to the crankshaft. There were several types of these engines designed. We have seen the crosshead engine called a steeple. There was a horizontal version used on some early screw-driven vessels. These engines were not too popular on the West Coast for a variety of reasons although a number of them were used.

There were a number of engines developed where the cylinders were mounted above the crankshaft. As far as we know they were not used on the Coast.

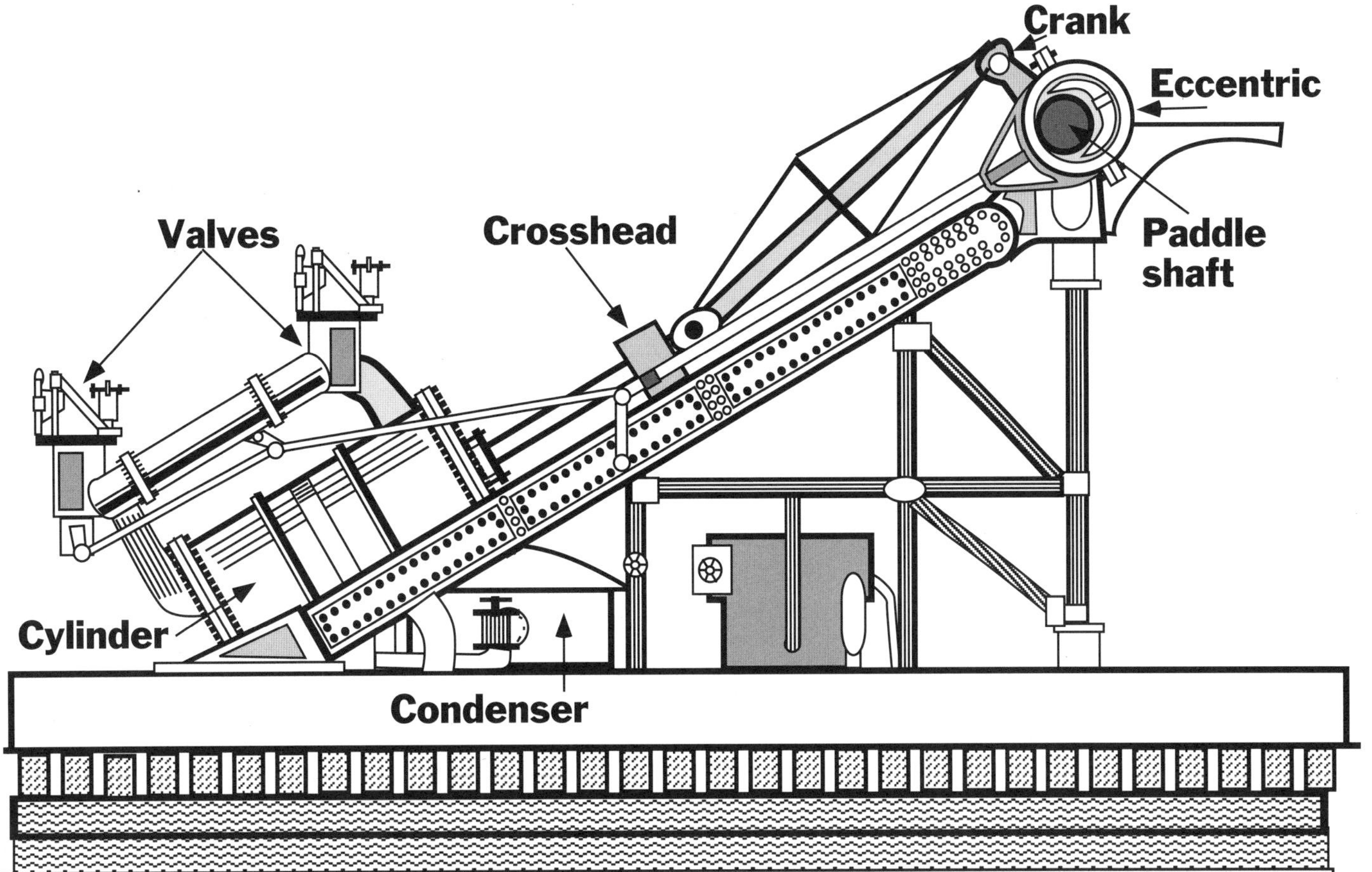

Low pressure inclined engine used on a lot of paddle wheel warships. Some were used in merchant ships as well.

Low Pressure Inclined

A number of U.S.N. ships had a low pressure inclined engine on their sidewheelers. One such vessel was the 181ft. (55m.) "sea going war steamer Fulton" built in 1852. Her displacement was 1108 tons. Her paddle wheels were 24 ft. in diameter. She had an inclined direct-acting engine supported on a wooden frame. The diameter of the cylinder was 50in. (1.27m.) stroke 10ft. (3m.). Originally she had a drop flue boiler which was replaced in 1857 with a vertical water tube boiler.

These engines had a large cylinder located as low as possible on the frame. Poppet valves were regulated by eccentrics in the same manner as other low pressure engines. The crosshead worked up and down the frame in guides. They were a lower engine than the walking beam. The Navy needed to keep the machinery as low as possible. There were a few of these engines used on passenger vessels operating out of San Francisco.

High Pressure Engines

Another type of paddle wheel engine used on the rivers and inland waters was the horizontal high pressure engine. They were installed at an angle to reach the paddle shaft. If two engines were used they usually operated each paddle separately. If one engine was used the engine would operate both wheels. For more information on these engines, see the chapter on sternwheelers.

The Inclined or Diagonal Direct-Acting Engine

The inclined-acting engine, or diagonal type paddle engine, replaced the vertical beam engine and the side-lever engine and in some cases the oscillating engine. This could not be accomplished until the quality of the metals improved to the point were the pistons could be laid on their side without distortion. Higher pressure boilers were necessary in order to keep the size of the cylinders down.

In Europe, a number of cross channel paddlers used one type or other of this engine. They were used extensively in North America as well, usually on ferries. There was at least one sternwheeler that used these engines.

This engine was a reciprocating engine, operating at an inclined angle. The heavy pistons were located below the paddle wheel shaft. They retained the basic steam engine principles. The steam was generated in boilers, stored in a steam chest and supplied to the engines as required. The engine direction was controlled by eccentrics as with other steam engines. The controls were simpler than the large beam engines.

Over the years, they ranged from simple double-acting single cylinder engines to large triple expansion and quadruple engines. They generally had two cranks but three could be used as well. The slide valves that control the steam entering the cylinders, were operated by the eccentric rod as with other engines. The valves could be of the piston type or the flat double-ported type.

On a number of the fast, narrow European vessels, some of the cylinders could be inclined (LP) while others were horizontal (HP.) The cylinders could also be located on opposite sides of the paddle shaft. In North America they were generally used on ferries which had more beam. These engines usually had the cylinders in line.

An example of the size of these engines was the Brighton Queen built in 1897. She had compound engines with the HP cylinder 30.5 in. (77.5cm.) in diameter and the LP cylinder measuring 65 in. (165.1cm.). The stroke was 5ft. (1.5m.). She operated at 45 rpm. with 140 lb. pressure. This gave the vessel a speed of 18.5 knots. Not a bad speed in anyone's books.

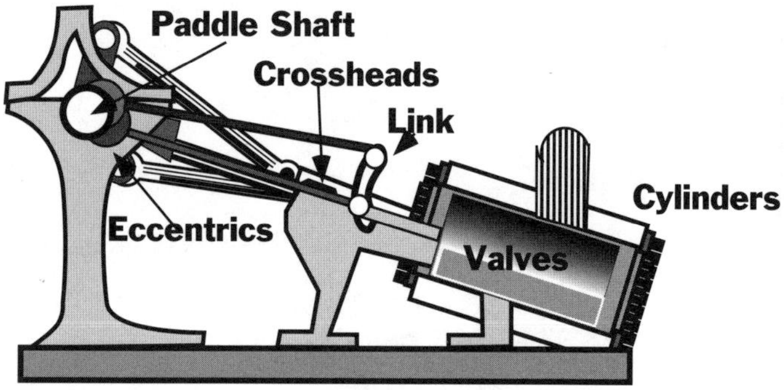

The webs and shaft of an inclined engine. The eccentric straps would be fitted to the eccentric bearings at the lower right.

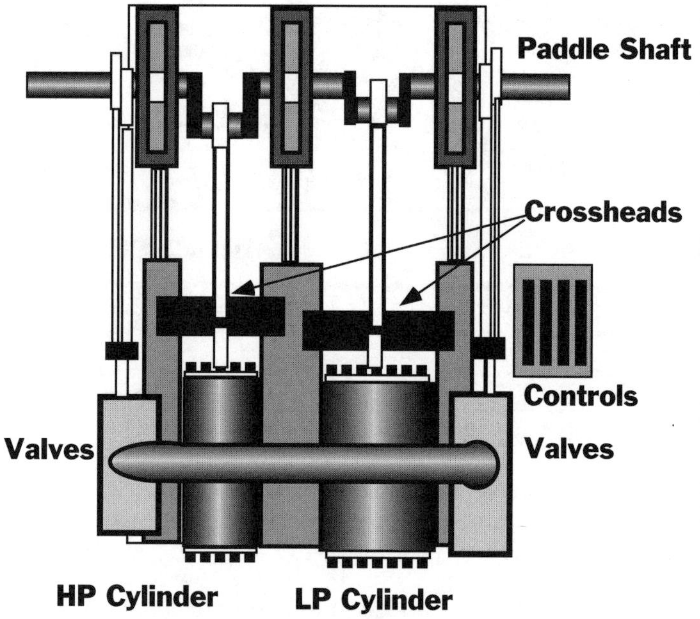

The inclined-action engine as might be found on a smaller ship. They could be triple or even quadruple expansion engines. There were some compound steeple types installed at an angle.
They often installed them side by side.

Vertical Direct-Acting Engines

In order to get away from side-levers, vertical direct-acting engines were developed in the 1830's. These were not the vertical direct-acting engine used with screw-driven vessels later on. They were not used for very long. They weighed less than the heavy side-lever engines. They were also shorter. The two cylinders usually were placed vertically under the centerline of the paddle shaft. The condensers and air pumps were located between them. The crankshaft was on a framework supported by the cylinders.

The connecting rods were very short which created side thrusts at the crosshead. The turning action was not as even as with the side-lever. To over-come this, some were made with open topped cylinders the connecting rod attached close to the piston with no crosshead.

There were a number of engines developed where the cylinders were mounted under the crankshaft. The siamese had two cylinders operating one connecting rod and one crank. A crosshead was used between the cylinders. A later version had the cylinders ahead and astern of the shaft. The piston rods were connected by a T-shaped crosshead.

Oscillating engine

This engine, despite much opposition, eventually became popular in paddle ships. It was also used in some early screw ships along with gearing to speed up the shaft speed. Patented in 1827 by Joseph Maudslay though originally proposed in 1785 by William Murdoch. It did not become popular until 1838. The first ones produced in the U.S. were for the "Republic" in 1849, made by Murray and Hazelhurst. These were low pressure engines. The later ones were compounding and operated at much higher pressures. The high pressure compounding engines were not popular on the West Coast.

The cylinders were not fixed as with most engines but oscillated on their centers eliminating the need for connecting rods. The cylinders could be placed directly underneath the crankshaft or at an angle to reduce the engine height. Each cylinder was mounted on trunnions. The trunnions were

The Oscillating Engine

On an oscillating engine the cylinders osculate back and forth doing away with the necessity of connecting rods. There were a number of designs over the years. The most successful were compounding with the cylinders side by side. The linkage for the valves was complicated.

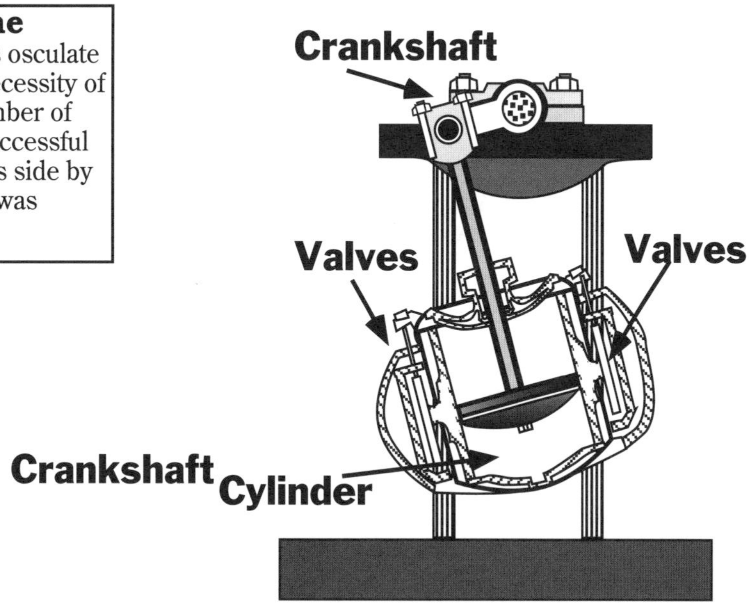

The Stephenson Link was able to work the valves despite the rocking of the cylinder. The die block connected to the frame is located in the link slot. The slot holds a second die block linked to the valves on the side of the cylinder. For purposes of clarity we have not shown all the parts. Some of these engines used levers to operate pumps and valves. Some used a smaller crank on the shaft to operate the air pump. Others used eccentrics to operate all or most of the equipment.

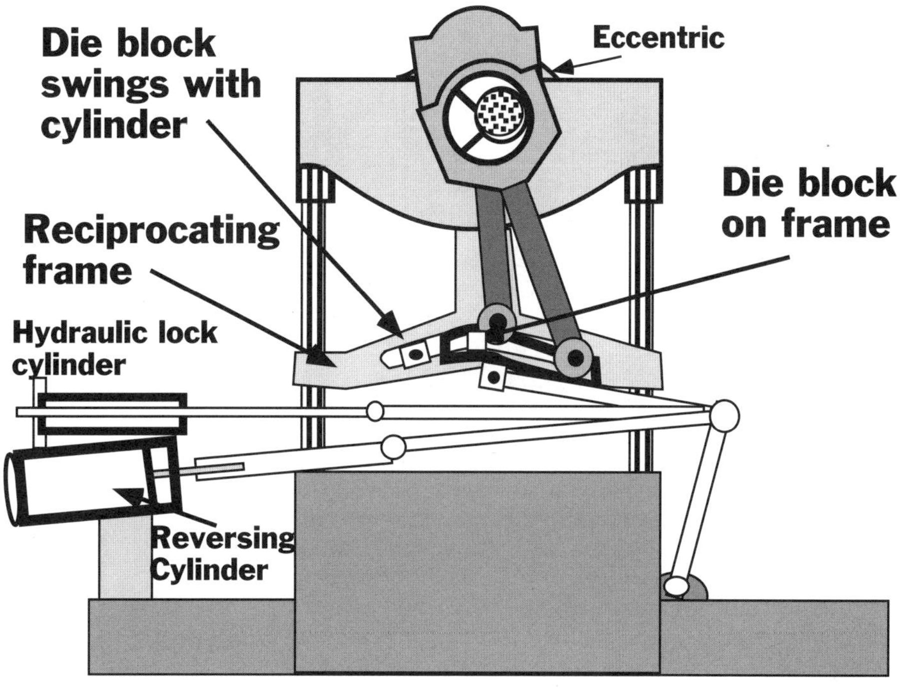

placed in the middle of the cylinder so they could swing or oscillate. Steam was led in and out of the cylinders by hollow trunnions. Valves located on the cylinders were operated by eccentrics. Compactness, few working parts, lightness and a low center of gravity were advantages. They often had problems with steam leakage at trunnion joints especially in the earlier years. Engineers had to constantly check the trunnion bearings for wear to prevent this.

Some had a condenser in the middle while others had them on the cylinders. On a number of these engines the air pump was operated by a smaller crank on the same shaft. The feed and bilge pumps were worked by large eccentrics fixed either on the trunnions or on the crank shaft.

The earliest ones produced by Joseph Maudslay had the cylinders directly beneath the shaft. The valve chest was alongside the cylinder at the trunnion giving a long passage to the cylinder ends. In 1838, John Penn placed the valve chests on the cylinders themselves, devising a linkage for driving the valves from the eccentrics without introducing error due to the osculation.

This made the engine viable. The engine was narrower and more efficient than the older ones. They were still not compounding engines at this stage of their development.

Over the years the engine went through a number of design changes. The number of cylinders and the angle to the crankshaft varied. One type of oscillating engine had three cylinders. One located directly under the shaft and the other two fore and aft of the shaft at a 60 degree angle. They were double-acting and connected to the same crankshaft.

Top - One of the later styles of osculllating engine. The valves are on the cylinder. The condenser would be separate from the cylinder on a number of models.

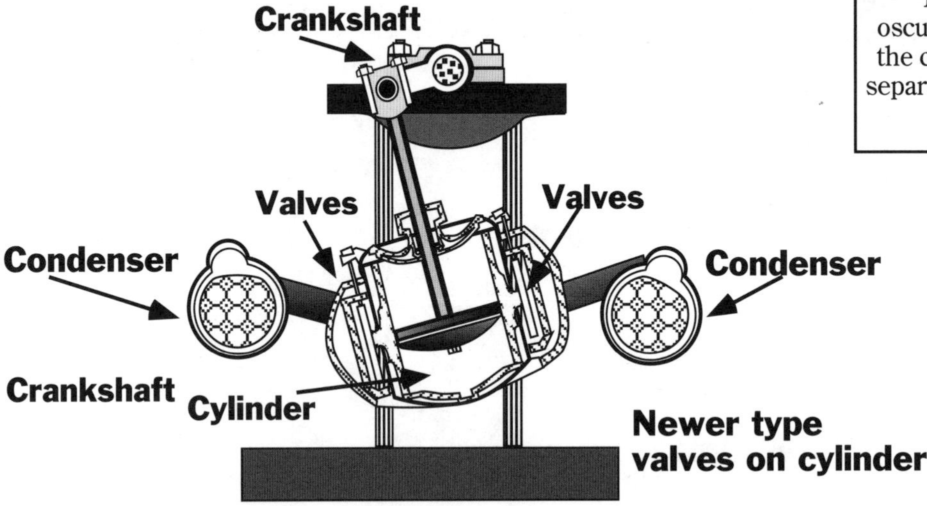

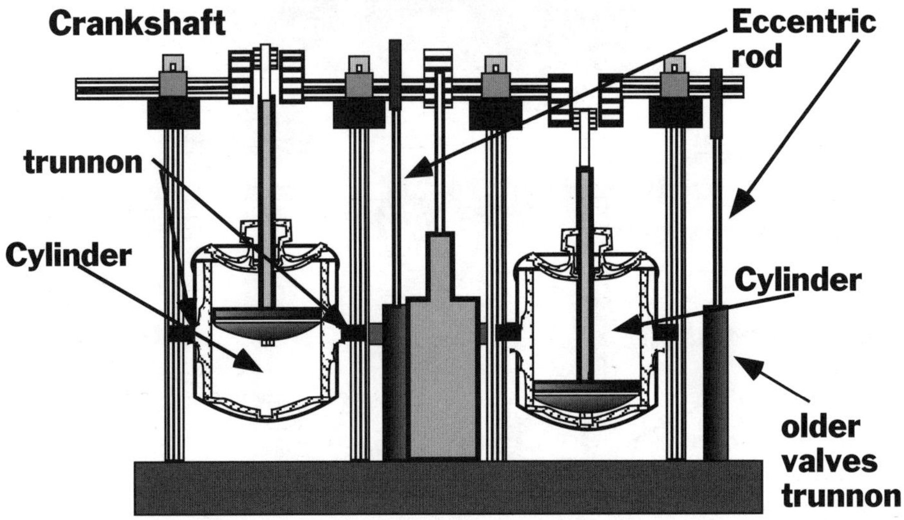

The U.S. built some of these engines for use on ships. Some of the American engines had quite a different design having two cylinders on one crank. There were major problems with these early designs that took some years to overcome. They did not become as popular in North America as in Europe. There were not too many of them used on the West Coast.

The American vessel the "Republic" worked out of San Francisco 1850 to 1864. Her osculating engines had 4.5 ft. (1.37 m.) pistons with a 6 ft. (1.83 m.) stroke.

They were quite a successful engine. Compound oscillating engines were still being built near the end of the 19th century. In 1860 the 343ft. (104.5m.), 2000 ton, PS Leinster was fitted with an oscillating engine which drove it at 18 knots. The two double-acting cylinders had a 98 in. (2.49 m.) bore and a 78in. (1.98 m.) stroke. She had an IHP of 4750. The boiler operated at 20 lb. psi. The paddle wheels were 32 ft. (9.8 m.) in diameter, revolving at 25.5 rpm. This vessel operated in England.

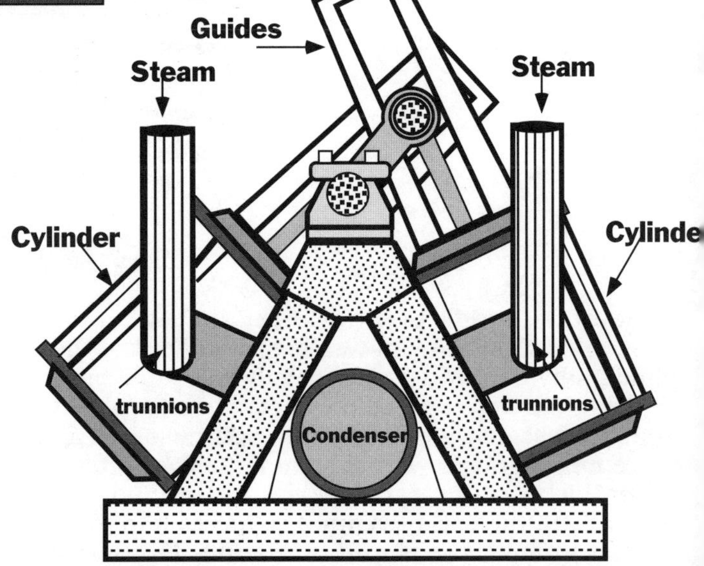

In the U.S. they used guides to cut down on some of the wear. They were a low pressure engine operating at around 30 psi. They used poppet valves and often used a starting lever. This engine was not as popular in North America as in England.

Sternwheelers
Chapter 5

Samson V
1937 to 1980
115 ft.(35m.) 418.4 GT
This sternwheeler was used as a "stump puller" on the lower Fraser River. She kept the river safe for shipping by clearing deadheads and other debris out of the river as her predecessors had done. Snag boats were used wherever there was shallow water. Flat shallow draft scow type hulls. Modern versions are used all over the Coast.
2 horizontal engines 63in X 14in (160cm x 35.6cm).
New Westminster, B.C.

Portland
Tug 1947 to 1981
219 ft.(66.8m.) 1800 HP
A tug used for taking ships through the many bridges on the Willamett River and the Columbia River. Similar to sternwheel tugs used on the Columbia for years.Steel hull (no king posts or hog chains).
Oregon Maritime Center and Museum
Portland, Oregon.

Moyie
1898 to 1957
161.7 ft.(49.3m.) 834 Gt.
Passenger vessel and tug. Operated on Kootenay Lake moving passengers and freight. In later years she handled barges.
2 horizontal non-condensing engines. 16in x 72in. (40.6cm x 183cm) 450 indicated HP.
National Historic Site, Kaslo B.C.

The Sternwheel Engine

There were more sternwheelers used on the west side of the Rocky Mountains than anywhere else in North America. They were used from California to the Yukon River.

Paddlewheel boats were used on the West Coast rivers and lakes soon after the first influx of settlers. First in the California gold rush then the Columbia River and soon after in British Columbia. The very first ones were sidewheelers but they did not work out too well in the fast rivers of the west. Sternwheelers soon proved to be the best option.

On the fast water rivers of the west the ability of the vessel to operate while going astern (backing) was extremely important. The sternwheeler with its multiple rudder system was very good at this. One skipper we talked to said they handled better going astern than ahead.

The boats varied from very crude small river boats to luxurious passenger vessels used on the lakes and inlets. Some worked as tugs pushing or towing small barges and others were freighters. These boats had to operate under all sorts of conditions including shoals, white water rapids, narrows and coastal swells.

The sophistication of the machinery was as different as the boats and their owners. Engines on the early boats came from all types of sources including some nonmarine.

Some of the first paddle boats used the same engines, boilers and hulls as Mississippi boats, others used walking beam engines. Eventually a West Coast version of the horizontal engine developed. The speed an engine could be manoeuvred was critical in fast water. The use of eccentrics and slide valves made for a faster operation than an engine that needed a starting lever and a "hook on" system for valve operation.

These engines were known as high pressure horizontal engines. At first they operated at a much higher pressure than the low pressure engines of the day. As pressures increased on ocean-going vessels this distinction ceased. Most of them were noncondensing engines. There were two types of these engines. The original western rivers type used poppet valves and levers, sometimes called a lever engine, and the slide valve engine found on the West Coast.

They were not the same as the horizontal engine used on early screw-driven ships. Some however were adapted for use on some small propeller-driven boats. There were many variations of these engines in use. However they had a lot of things in common inherited from their roots in the western river system.

W.T. Preston
1929 to 1981 A snag boat operated by the U.S. Army Corps of engineers. Originally a wood hull. 2 150 IHP engines built in 1914. Length 163 ft. 6 in. (49.83 m.) Displacement 494 long tons.
Anacortes Museum, Anacortes Washington.

Sicamous
Lake boat - Lake Okanagan- 1914 to 1935. Length 200 ft. (61 m.) 2 compound engines 101.3 NHP. Jet condenser. 1786.7 tons.
The lake boats usually were larger, slightly deeper and had a finer entrance angle (sharper bow) than their river boat cousins.
Pentiction B.C.

The Western River High Pressure Engine

As far as we can tell the first high pressure, non-condensing horizontal engine was installed on the Washington in 1817 on the Mississippi River. Captain Henry M. Shreve and his associates are credited with its development. Whether it was a sternwheeler or a sidewheeler no one knows. Up to this time most of the paddlewheel engines operated just above atmospheric pressure. The horizontal engines operated with pressures of 50 to 150 psi. These pressures caused some major problems with the early boilers. There were many boiler explosions until improvements were made.

The low pressure engine did not have enough power for river work. A high pressure usually non-condensing engine using 40 psi was built for the rivers. By 1840 pressures of 100 psi had been used. The cylinder was far smaller in diameter than the low pressure engine. They were also much lighter a very important consideration on a shallow draft river boat. They were cheaper to build as well. The first ones were vertical but they soon became horizontal.

Judging from drawings of 1816 the first high pressure engines did not look much like the later types of horizontal engine. Most of the engines drove sidewheel boats though a number were used on sternwheelers. The engine as we know it was developed by 1830-40.

The vessels that used these engines were very limber. They had to go over sandbars and shallow places without straining the hull. With all the stops made up and down the river the loads were not distributed evenly in the hull. This meant the engine had to operate and also allow for movement. This is the reason for the use of levers and cams.

There were two basic types of engines used with many variations. The difference was in the valves, one having fixed cutoff and the other variable cutoff valves. As levers were used to operate the valves they were sometimes called a lever engine.

If the engine was used by itself it would have a flywheel to make it run smoother. The paddlewheel served this function if two engines were used. On sidewheelers the engine would be mounted on a slant.

The cylinder was relatively small in diameter with a long stroke. An example of one engine is a 24in. (61cm.) diameter cylinder with a 96 in. (2.43m.) stroke. The piston drove a long connecting arm known as a pitman to a crank on the paddle shaft. The valves used were of the poppet type. The steam valves with their connecting pipes are on one side of the cylinder and the exhaust valves on the other side. Each set of valves were operated by separate rocking cams or levers which got their motion from rockers and connectors from a special cam on the main shaft.

The cam worked very well under a variety of river conditions. The twisting of the boat when loading and unloading could be accommodated.

The valve operation could be changed to allow for the existing conditions on the river. Commonly two calms were used, a full stroke cam and a cut-off cam. When the full stroke cam was used, all four valves were used raising one exhaust and one receiving valve at opposite ends of the cylinder at the same moment. This was reversed as the piston moved in the opposite direction. The same cam was used for ahead and astern. After the boat was making way the cut-off cam could be engaged. This valve cut off the steam before the end of the stroke. It could only be used in the ahead mode.

As with all steam engines there were a lot of variations in linkages. Some used the Sweeney valve gear while others used Rees variable cutoff, a California cutoff, or a two valve engine. They all used a cam rather than eccentrics. Some had a form of slide valve combined with a poppet.

The engine usually exhausted directly to atmosphere and not back to the stack. The engine base and pitman arms were often made of wood to give them more flexibility.

On sternwheelers the controls for both engines were connected for ease of operation. On sidewheelers the engines and their controls were usually separate.

These vessels often used a "doctor" which was a feed pump, fire pump and feed water heater run by a small vertical beam engine. As they would work with sand in the river water, they could be used when injectors could not.

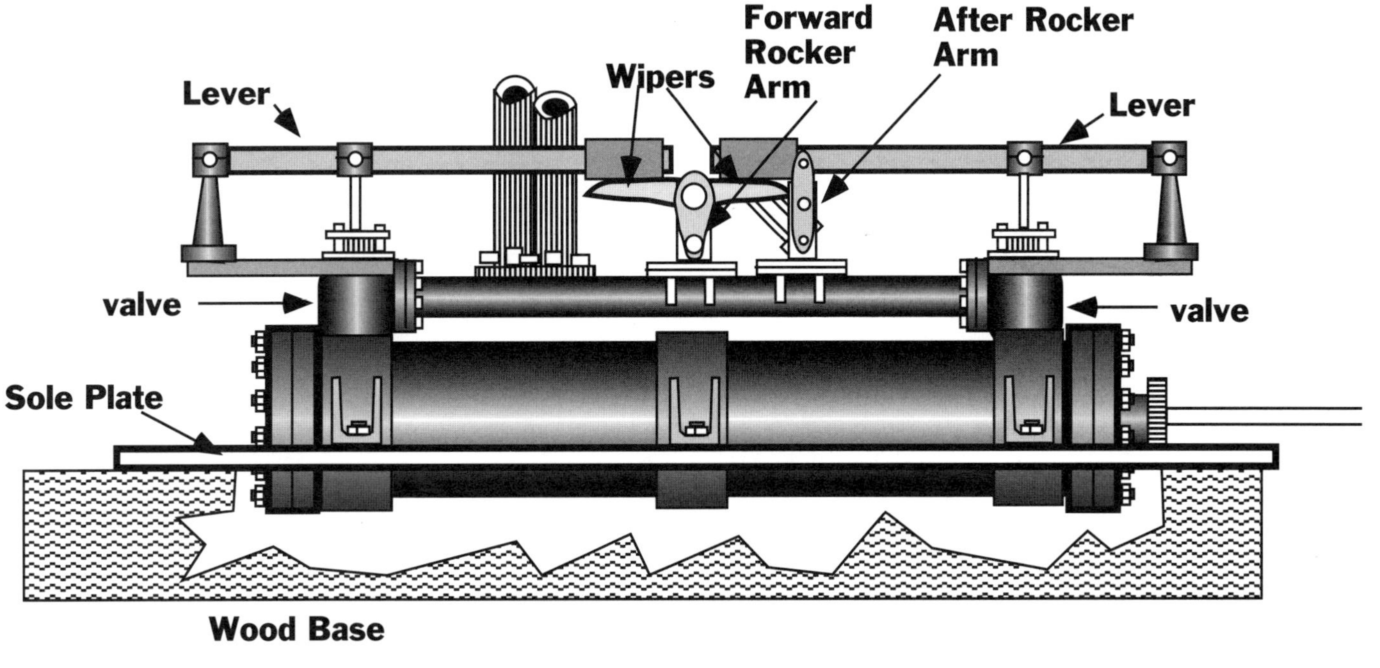

A western rivers-type fixed cut-off engine.
There was a complete set of valves and levers on each side of the engine. One set was for steam (port side) and the other side was for the exhaust (starboard side). The valves were operated by levers and wipers. The wipers got their motion from the rock shafts. When under way the rock shafts were hooked on to rods going back to cams on the crank shaft. Some engines had other methods for using cut-off cams but the principle was the same. The cams used worked well when the hull moved under conditions of stress. They were better than eccentrics under these conditions.

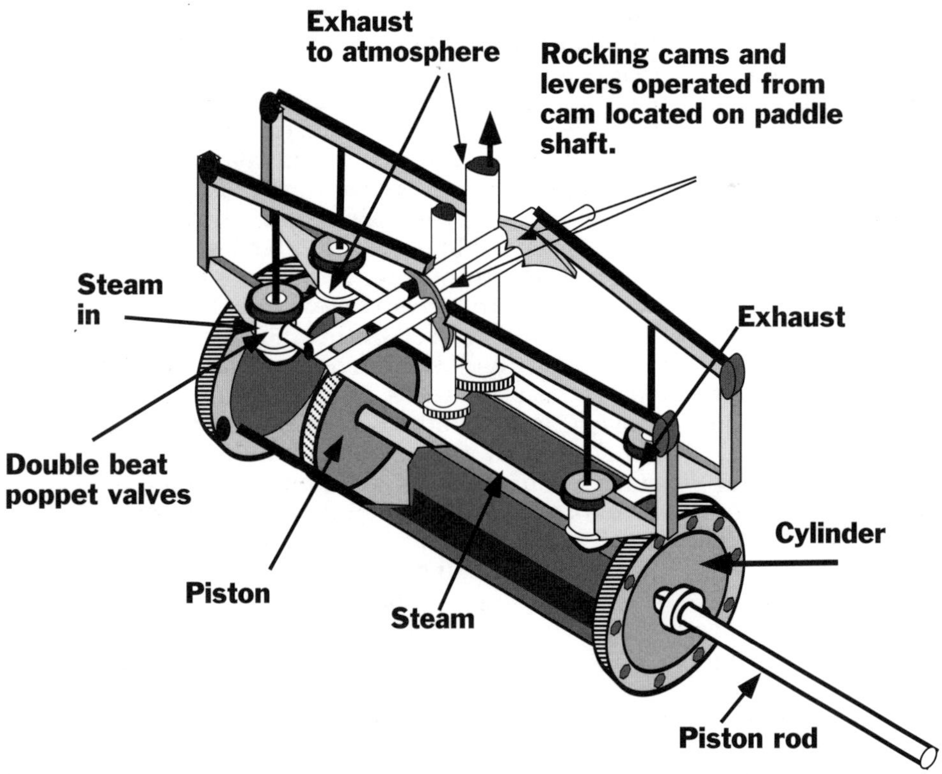

The horizontal engines used on the Mississippi river boats used poppet valves rather than sliding valves. The engine had to use a starting lever (not shown). The valves were operated by cams on the paddle shaft. In order to reduce clutter in the illustration we did not show pumps and other parts. These engines were also used on some sidewheelers.

Fixed cutoff Western River Engine.
The piston rod moved the crosshead fore and aft. This in turn operated the long pitman arm connected to the paddle crank. The rods from the cams went back to the engine to operate the valves.

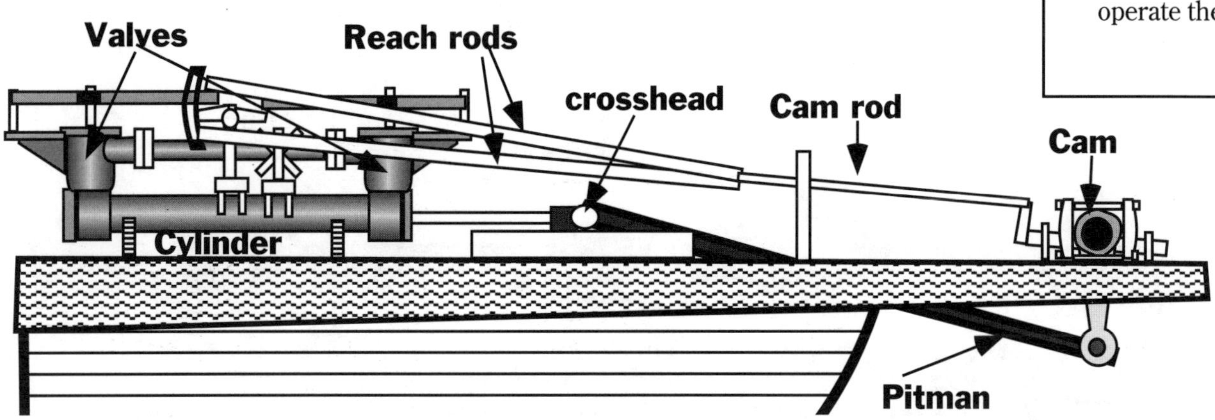

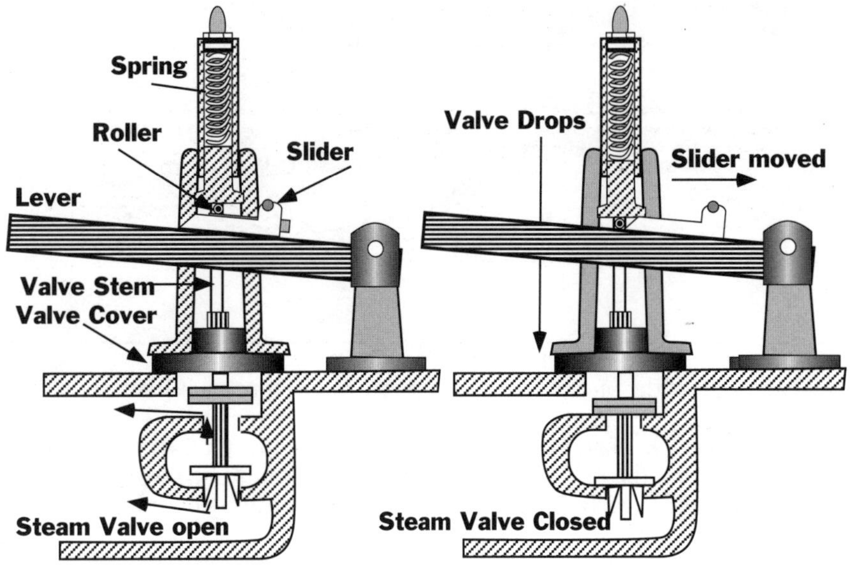

In order to overcome problems with cut-off valves some new valves were invented. The "California Cut-Off" put the valve under the pressure of a spring. The lever opens the valve in the normal manner. When the slider (controlled by the cut-off cam) was over, it held the roller and the valve open. When it was moved back the roller and the valve dropped, closing the steam valve. This and other systems were quite complicated. A West Coast invention that went East.

Horizontal Steam Engine On The West Coast

The fast rivers and open ocean operations of some of the West Coast sternwheelers brought a different set of problems to be overcome. The engines used had a lot of the characteristics of the Mississippi boats but the valves and controls were different. To get better draught they sent the exhausted steam back to the stack as was the practise on the railroads.

The engine that eventually was used in the west was mounted on heavy metal beds on either side of the ship, running from the forward part of the engine to the after end of the engine room. The long horizontal cylinder contained a double-acting piston. The piston rod was connected to a crosshead that moved backwards and forwards along guides secured to the engine bed. The crosshead moved the pitman arm connected to the crank on the paddle wheel. The cranks were set at different angles on opposite sides of the ship so the engines would not get centered when manoeuvering. Locally the pitman arms have been called walking beams due to their back and forth movement but this was an error.

As with all steam engines valves were required to regulate steam entering and exhausting the cylinder. This in turn determines the speed and direction of the paddle wheel. One or more eccentrics were located on each side of the paddle wheel. Long rods ran forward to a linkage connected to the valve shaft. The use of slide valves and eccentrics made these engines fast handling for West Coast conditions.

The valve system varies with the type and complexity of the engine. Some of the smaller river boats just had a single "D" valve in a steam box on the center of the cylinder. Most of the engines had valves on each end of the cylinder, operated by a common valve shaft. Some of the engines had cut-

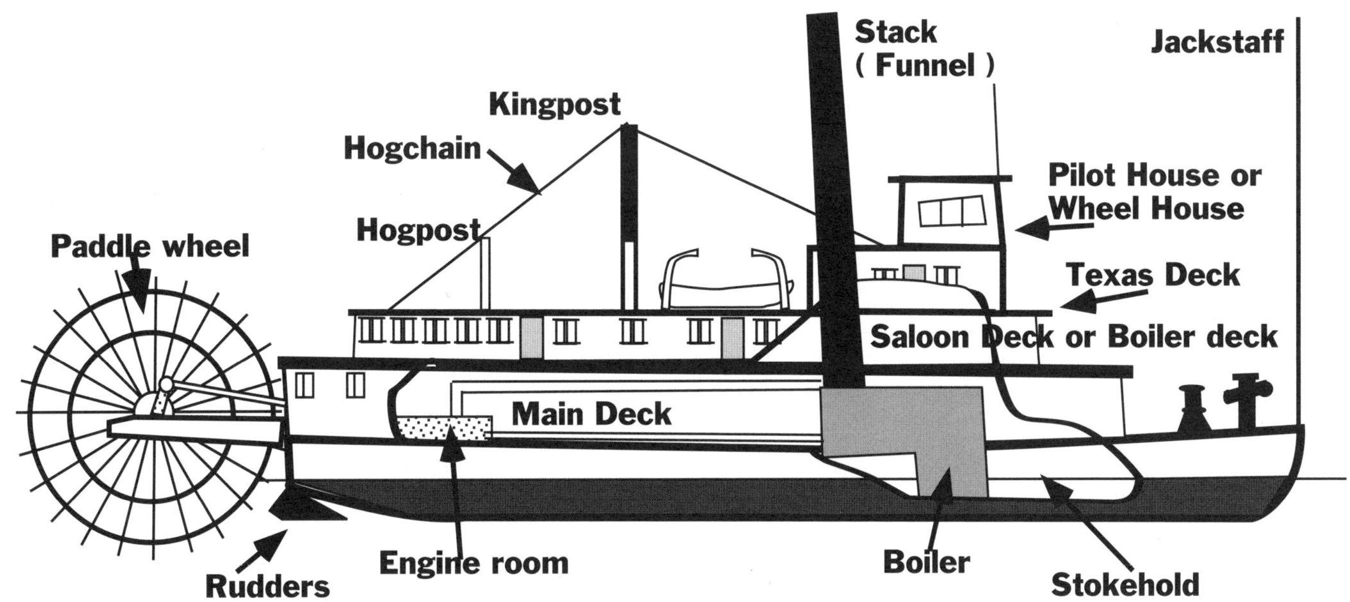

The design of stern wheelers and their engine rooms followed a basic pattern. To balance the weight on these very shallow draft vessels the boiler and bunkers were always forward and the engines and wheel aft. The actual vessels were different in size, boilers, engines and number of decks.

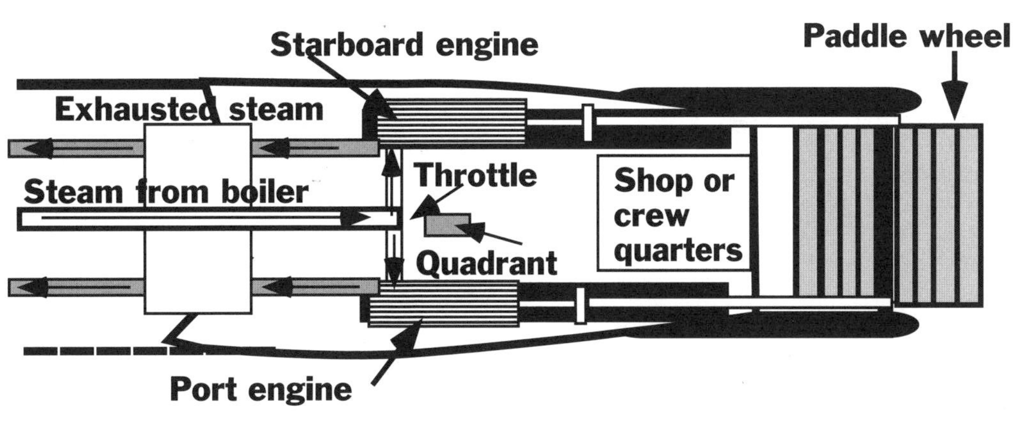

The engine room of the stern wheeler Moyie.The engine rooms on these ships always seem spacious when compared to coastal freighters. The controls are in the center engines on each side.Steam gages are on the bulkhead near the controls. Pumps and the dynamo (electrical generator) can be seen in the foreground.

off valves that could be controlled from the quadrant. This would give the engineer greater control of the paddle wheel. When a steam engine was running efficiently the exhaust valve should be opened for most of the stroke. The steam valve should only be opened for part of the stroke letting the expansion of the steam do the work. On some engines a cut-off valve is used to shut the steam off at the right time. This cut-off valve could be regulated for the type of work the ship was doing. On the tugs it was often not used as they needed all the power the engine could put out. Other vessels needed all the power they could get when working in fast water. Backing power was as important when coming down river as ahead power going up river.

Large compound engines like the one on the "Sicamous" a lake steamer, had a complicated system that included a cut-off valve. The linkage controlling these valves became more complex as well. Compound engines did not respond to manoeuvers well making them unpopular on the fast moving rivers. They worked well on lakes and in semi-protected waters such as Puget Sound.

The linkage was connected by rods to levers in the center of the engine room. The engineer could control the action of the paddle wheel with levers on a quadrant. On most engines the throttle was a valve on the main steam line while others

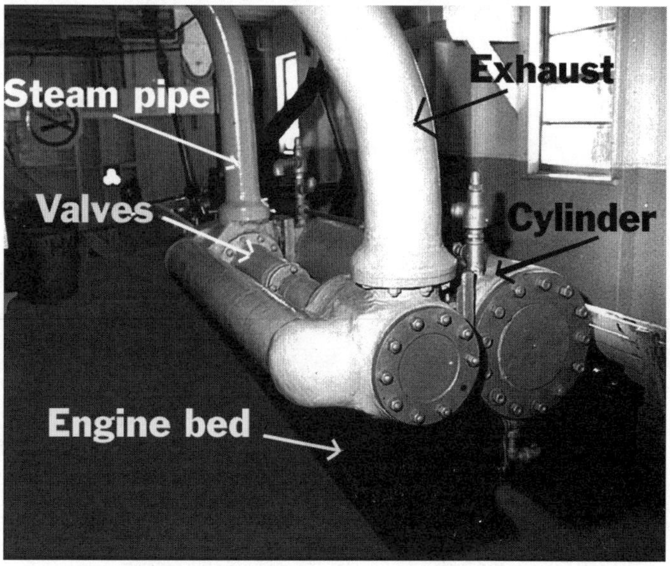

Port engine Samson V. This vessel had piston valves but no cutoff valves. All these engines were based on the same concept but individual in design. They all had a solid iron or steel base. They all used slide or piston valves.

The engine on the Moyie has cutoff valves controlled from the quadrant (control station).

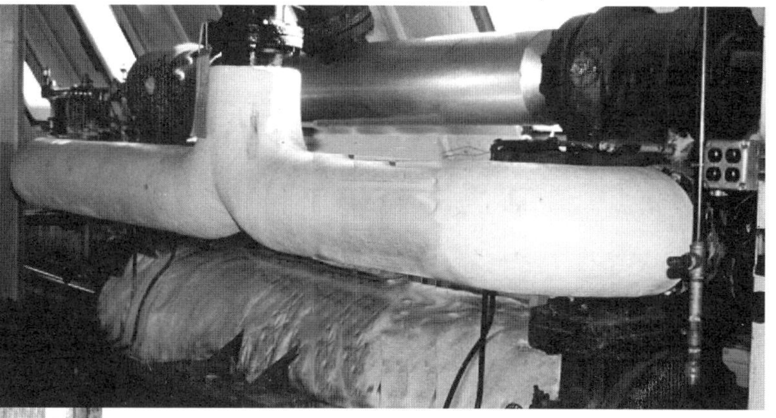

The engine on the tug Portland. Note covers on the main cylinder (lower part of photograph) to reduce heat loss due to radiation.

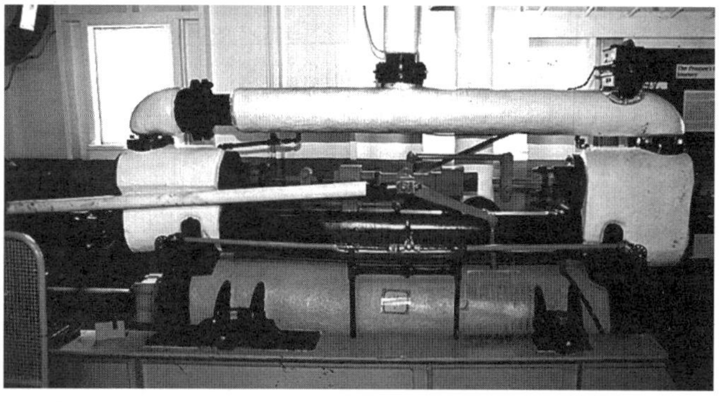

Engine of the snag boat W.T. Preston.150 nhp engine.The piston valves are operated from eccentrics but have a unique control system for the main valves and the cut-off valves.

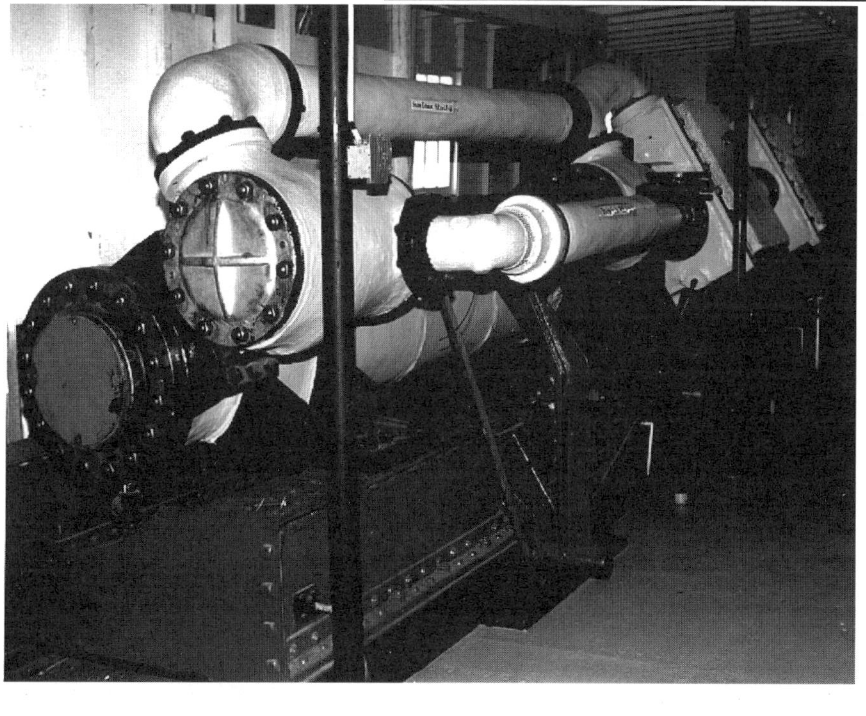

The Compound engine on the Sicamous. Compound engines were more fuel efficient. They were slow handling for river work but worked well on the ocean and lakes. This vessel had a jet condenser.

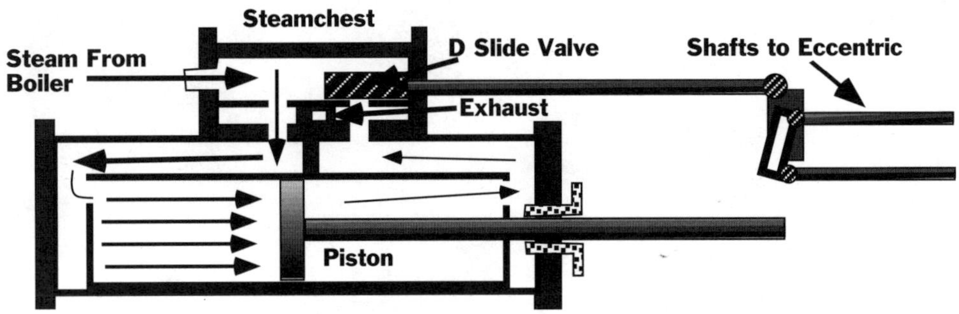

This type of engine with a single slide valve was used on some small river boats

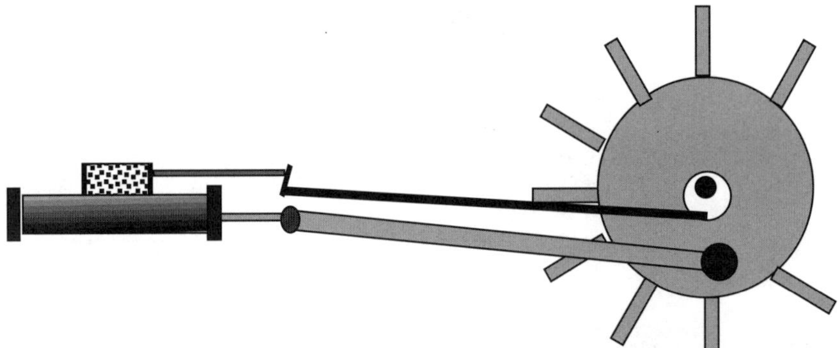

Basic engine without variable cut-off valves

This diagram shows the basic way these engines work. Some of these engines had another set of valves called cutoff valves.

Piston valves are shown here but slide valves could be used. Piston valves work better with higher pressures.

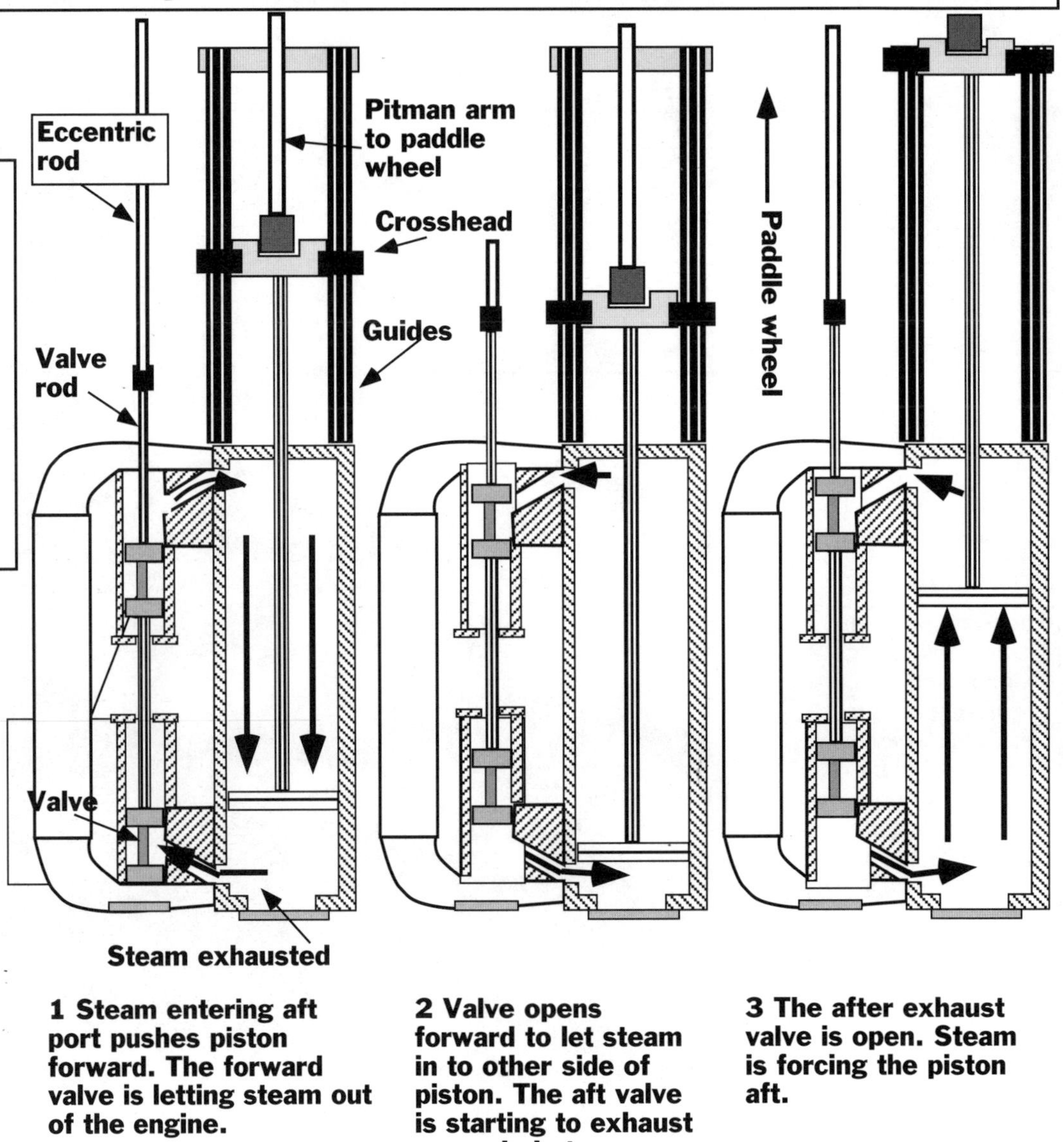

1 Steam entering aft port pushes piston forward. The forward valve is letting steam out of the engine.

2 Valve opens forward to let steam in to other side of piston. The aft valve is starting to exhaust expanded steam.

3 The after exhaust valve is open. Steam is forcing the piston aft.

The basic layout of an HP horizontal engine. To make the diagram easier to understand we have separated components. See photographs. The piston traveling back and forth in the cylinder moved the long pitman arm in a back and forth motion to turn the crank on the paddle wheel shaft. This back and forth motion caused the pitman arm to be misnamed a walking beam. The cranks on opposite sides of the wheel were usually set at 90° so the paddle could be operated ahead and astern without getting jammed.

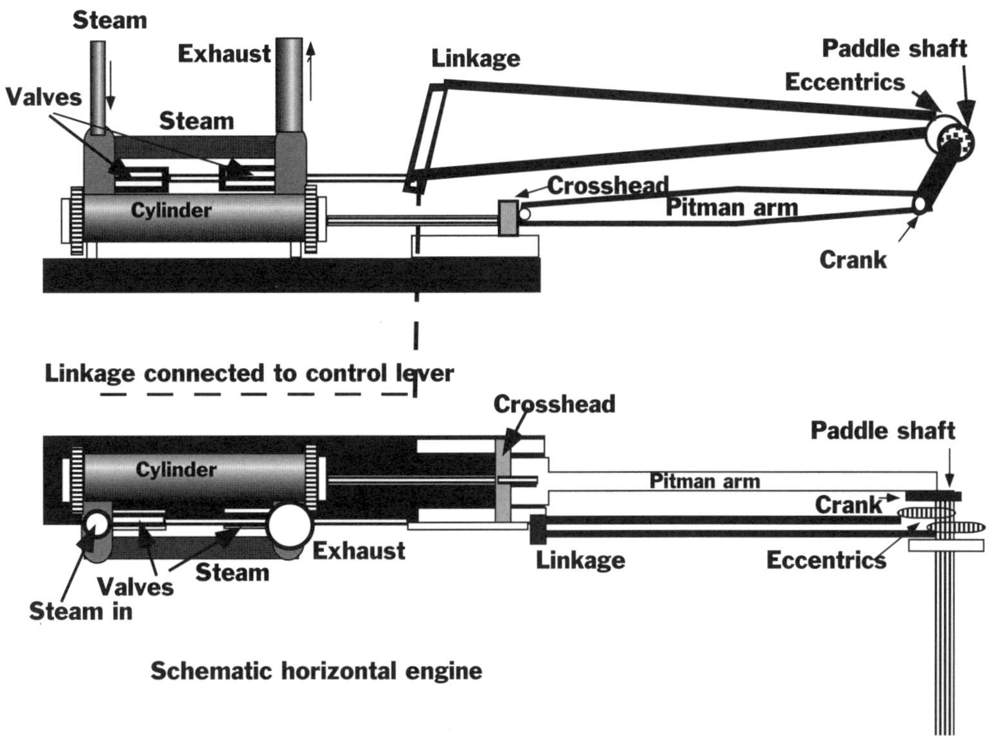

Schematic horizontal engine

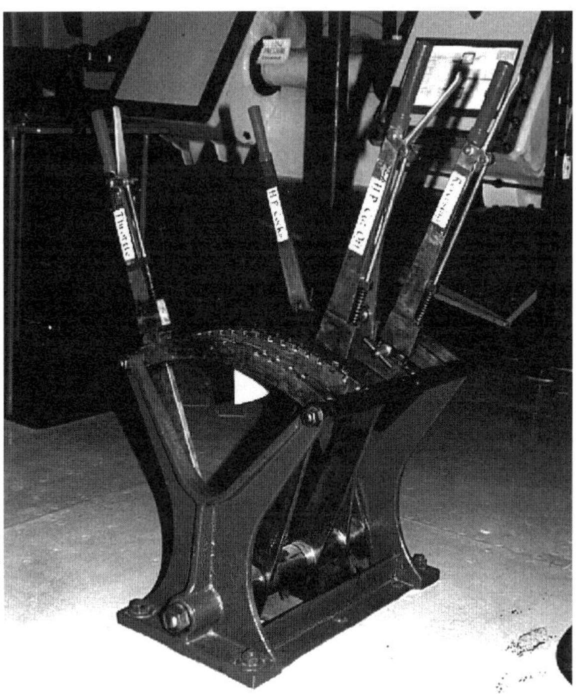

The quadrant controlling the compound engines on the Sicamous. This type of compound engine with cutoff valves needed a complicated system of levers and linkages.

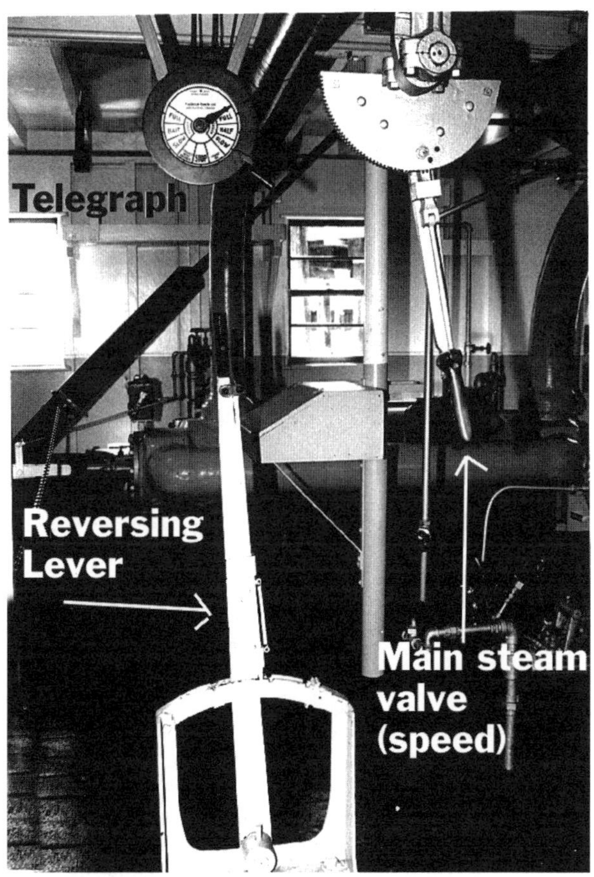

Control station on the Samson V. The reversing arm operated the linkage for eccentric.

Engines with variable cut-off valves

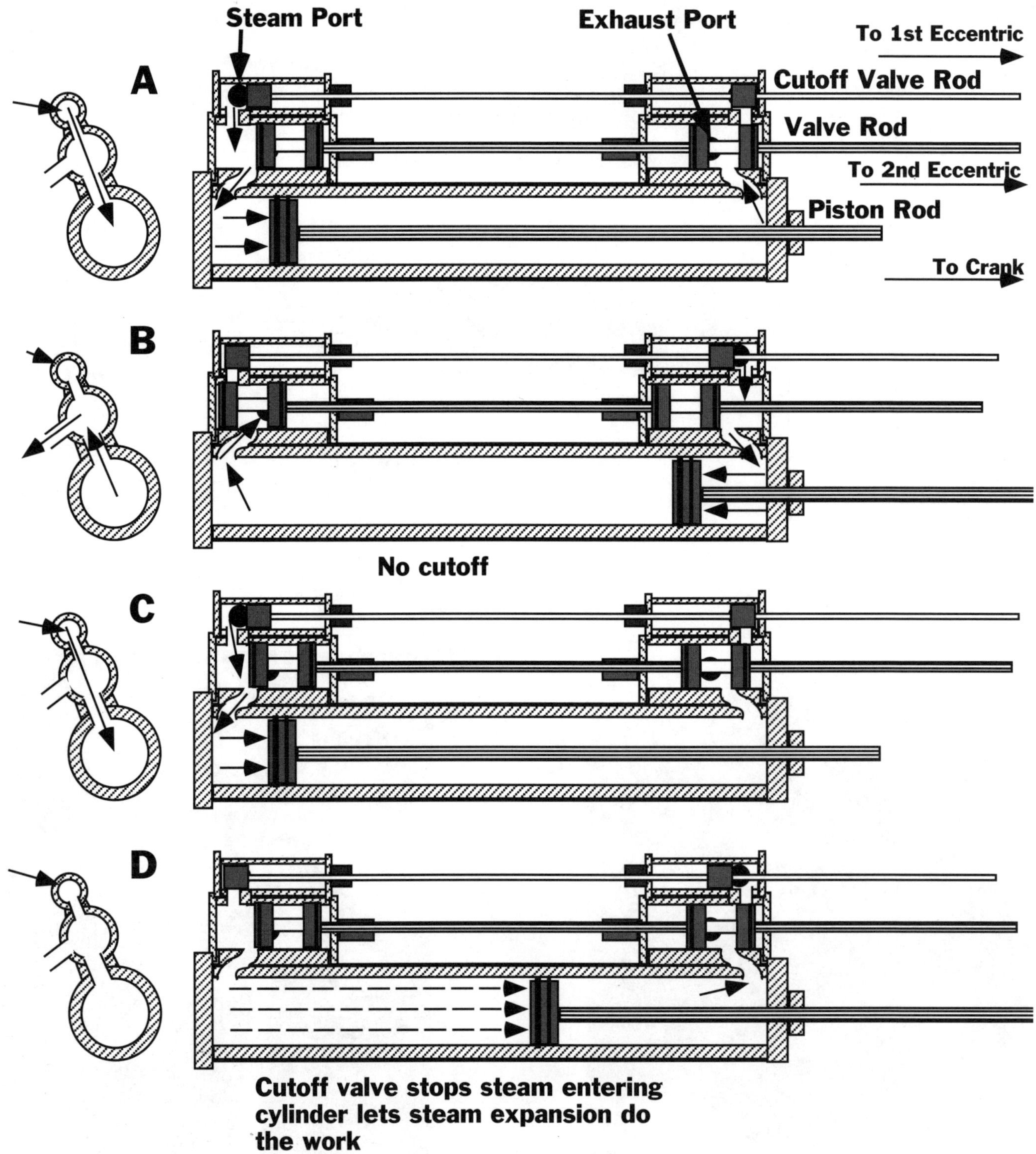

The use of cutoff valves.
As can be seen from the previous photographs each one of these engines had slightly different valves. One way of operating the cutoff valves was to use a separate eccentric and control lever to work the cutoff valves. At **A** and **B** the cut off is not in use. The valves induce steam from the boiler at one end and exhaust at the other. In **C** at the start of the stroke the steam enters as before. At **D** the cutoff valve shuts off the steam from the boiler. The expansion using the latent heat in the steam keeps moving the piston to the end of the stroke.These engines had a very long stroke so this system worked. In normal practice the cutoff valve would be too far away from the valve to be effective.
These vessels needed to control the cut off so they could utilize all the power available when working fast water, towing or manoeuvering. Smaller engines might use slide valves.

The usual type of eccentrics on the paddle shaft wheel operate the valves.
W.T. Preston

Links on the Samson V. The lever at the quadrant moves this linkage which controls which way the paddle turns.

had a lever connected to valves on the engine for this purpose. All these links and rods had adjusting screws so the engineers could keep two engines synchronized.

Steam was brought aft from the boiler by a large pipe located just under the saloon deck or boiler deck (one deck above the main deck). The pipe split just forward of the engines to supply steam to each engine. On some vessels the throttle valve and lever was located before the steam line split.

Steam was exhausted through pipes running under the main deck on a lot of vessels. In most cases the steam went straight back to the steam box in the boiler before going up the stack. This produced a lower exhaust pressure required for the engine to operate efficiently. Sending the exhausted steam, known as "a steam blast," up the stack to increase the rate of combustion was an old principle used on naval ships and rail road engines. Some ships did exhaust the steam to jet condensers.

These engines ran for years and often outlived the ship. In many cases the engines were taken out of old or wrecked vessels and installed in new hulls. On the other hand the boilers often did not last for the life of the vessel.

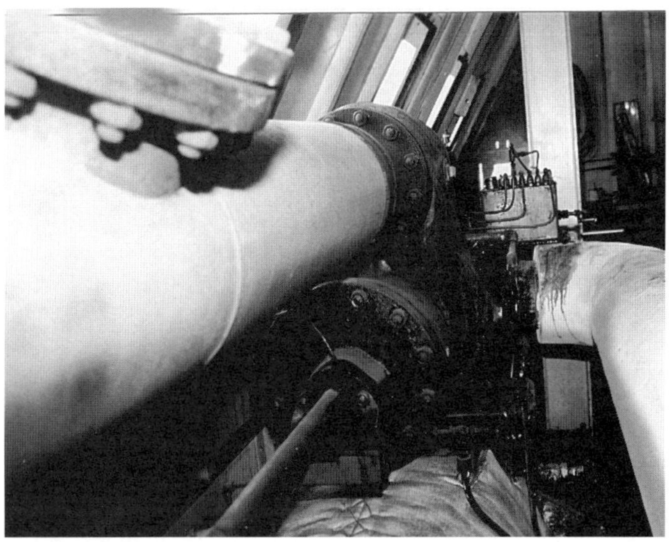

The stern wheel the tug Portland uses a cam rather than an eccentric to control the piston type valves, seen on the right. The cam runs off the pitman arm. This is an unusual system.

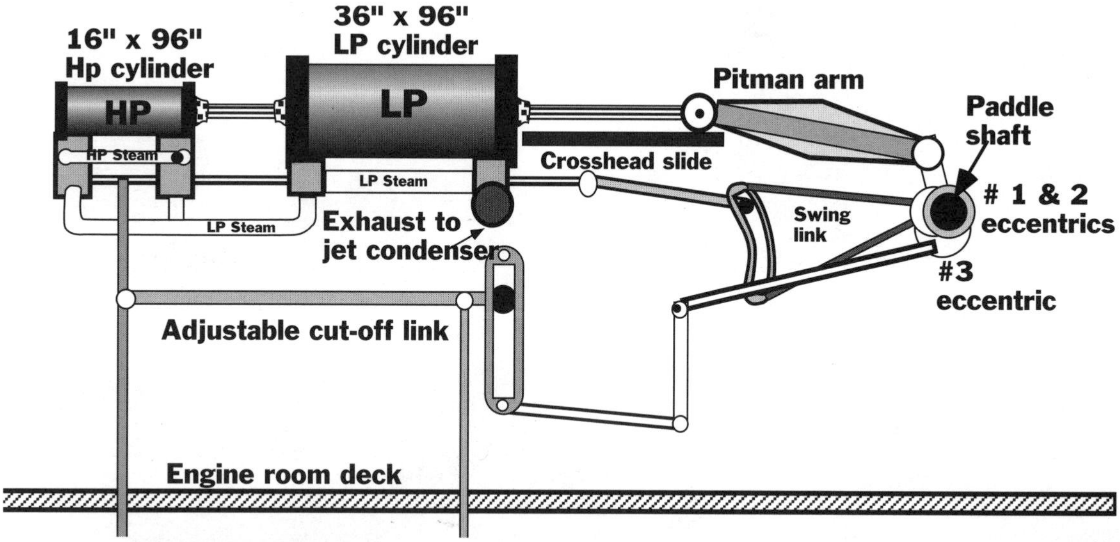

Diagram of the controls on a compound engine with cutoff valves. The linkage is quite complicated.

Crosshead on the Portland. The piston moves the crosshead up and down on sliders. This moves the pitman. Note the little oil plate that keeps the slider lubricated.

The pitman arm operates the crank on the paddle shaft. The cranks on each side are at 90 degrees to each other so there will be no problems starting the engines.

Crosshead on the Samson V.
Note rods were used for crosshead guides rather than flat plates.
(New Westminister B.C.)

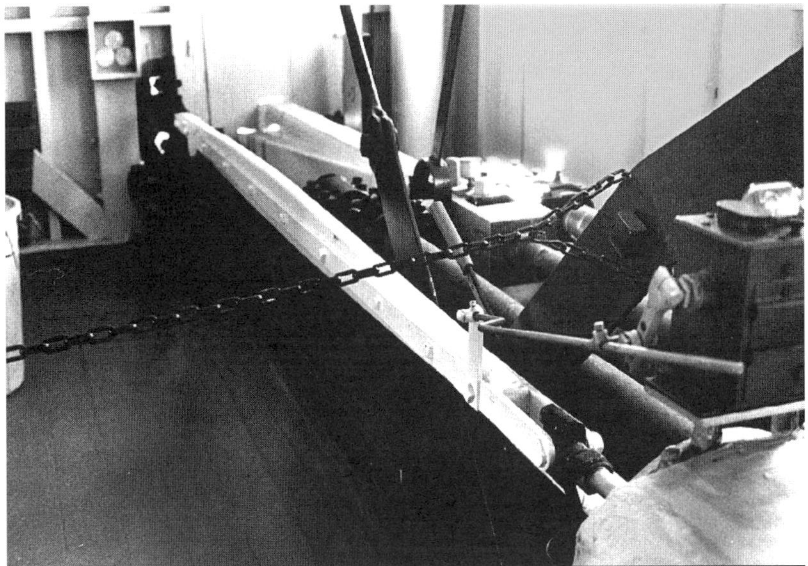

The crosshead, pitman arm and piston rod on the Samson V. The closest arm connects the eccentric to the valve rod.

On wooden hulled river boats the hull had to be kept from sagging by the use of hog-chains and hog-rods. This kept the machinery in line.
The kingposts and hog-rods of the Samson V.

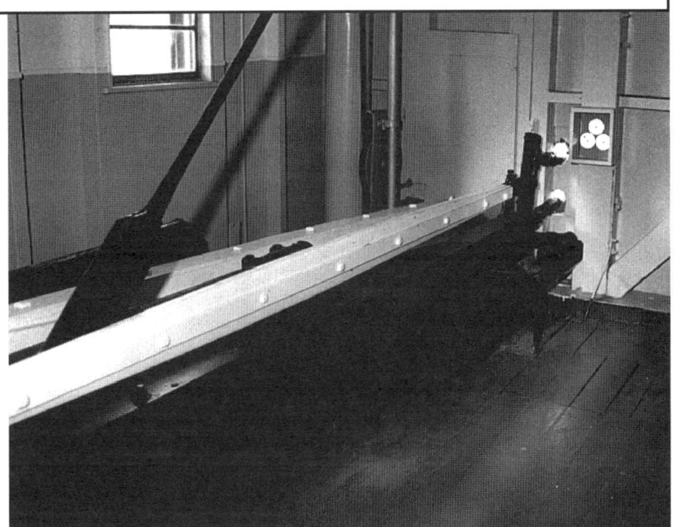

Pitman arm and links on the Samson V. When the vessel was operating the doors in the after end of the housework were opened up to let the arm work.

Staggered paddles on the W.T. Preston to reduce wake. (Anacortes Washington)

Stern wheel San Francisco Historic Park
This shows the size of the paddle wheel used by some of these vessels.

Screw-Driven Ships

Chapter 6

The Engines That Developed The West Coast

A triple expansion reciprocating steam engine commonly called an "up and downer". The engine was built in England in 1932. It powered a 175 ft. tanker on the Great Lakes. The condenser and pumps are not shown. Preserved at the Toronto Maritime Museum.

There were many attempts to use paddle wheel engines to operate screw driven ships. This model from the San Francisco Maritime Museum shows oscillating engines used with gearing to drive the propeller. Note the engines were still below the shaft. Gearing was used to increase rpm and supply power to the propeller shaft in the lower part of the hull.

Engines For Screw-Driven Vessels

Propeller or screw-driven ships had been around for years before they became reliable enough to replace paddlewheelers. When the screw was first introduced engines designed for slow turning paddle wheels were used. Due to manufacturing problems the propellers were small. In order to make the small screw effective the shaft had to operate at a far greater rpm. Large gears were tried to overcome this problem. Gearing and small screws existed even after technology allowed for the construction of larger propellers. This equipment was heavy and needed a lot of repairs. These problems extended the life of the paddlewheel ships for many years.

Contemporary literature has it that when the HMS Rattler and the HMS Alecto compared paddle wheels to screws in 1845, the screw-driven vessel proved to be superior in all cases. The Navy preferred propellers, as paddlewheels were vulnerable in battle and interfered with the guns. According to Mr. John Bourne C.E. an eminent engine designer, in his book "A Catechism of the Steam Engine" this was not totally true. He goes into a lot of the technical details of a series of tests done on these vessels and others but the main points are these:

1. For ocean vessels in calm weather or with side winds there was not much difference with light and medium immersions (loads).

2. Under similar conditions with deep immersions (heavy load) the screw was better.

3. In head winds under all conditions of load the paddle wheels had the advantage in fuel consumption and would be just as fast.

4. The best use of the screw is in auxiliary power for sailing ships.

5. For long voyages in areas such as the trades where wind could be expected sail was by far the best.

We must remember the engines and the propellers were not very well developed at this stage. When better hull design, better screw design, compounding engines and higher pressures came along this all changed. We noticed the large screw-driven vessels of the 1850-1860 period did not have as long a life as the paddlers of the era.

In the U.S. other problems had to be overcome. Iron ships were not built in North America until some years after they had been perfected in Europe. The larger wooden hulls were too flexible for the long propeller shafts. However smaller low powered coasters of under 200ft. (61m.) and tugs worked well. Putting the engines aft as with the steam schooners helped to overcome the problem. This is one of the reasons that while steam worked very well in the early days on the West Coast, sail was still king on the long ocean voyages. Many of the early ocean-going screw-driven ships were under powered and relied mainly on sail.

Metal stern bearings were used at first. The copper sheathing used on some wooden vessels created a problem with electrolysis in the stern bearing and the tail shaft. This was overcome by using lignum-vitae, a very hard wood for stern bearings.

A number of existing engines were tried including the steeple engine placed on its side and the oscillating engine. The first really successful engines, for screw-driven vessels were horizontal engines including the return connecting rod engine and forms of horizontal direct-acting engines. The horizontal trunk engine was used by the R.N. for a while. A vertical trunk engine was tried in the U.S. without the greatest success.

After rejection in the U.K. John Erickson went to the U.S. and built the first screw-driven vessel in the U.S. Navy. She was the Princeton built in 1843. Horizontal engines were used with great success on U.S. Navy gunboats during the Civil War.

Rotatory engines, the ancestor of the turbine, were tried but did not prove too practical at this time. The first direct-acting (no beam) vertical engines were paddle wheel engines. They had the cylinder under the crank shaft.

Eventually the vertical direct-acting engine

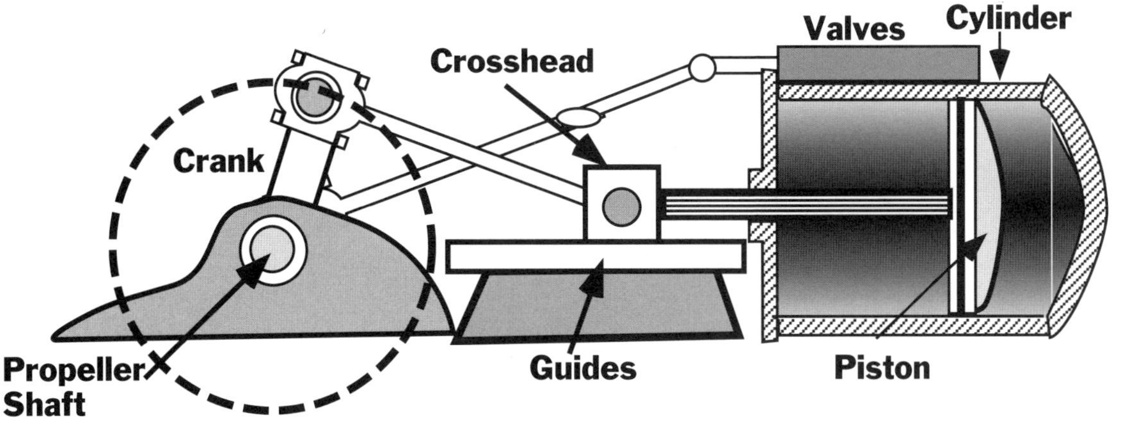

The Horizontal Direct-Drive Engine.
These engines had a short connecting rod as thwartships space was limited. These engines needed a large cylinder which gave some balancing problems.They operated with less than 30 psi. They were a simple engine and the forerunner of the vertical direct-acting engine. They were fitted on the beamier warships for some time. The Navy liked to have the engines below the water line. The diagram does not show all the engine.

Horizontal Engine

There were a number of horizontal engines in use on merchant ships and naval vessels between 1850 and 1860. A lot of ocean going ships were using these engines along with a screw. They operated with 20 to 25 psi. A number of them used jet condensers until around 1870. The navy used them longer than the merchant marine. The types most commonly used were the direct-acting engine, the trunk engine and the back-acting or return-connecting rod engine. When higher pressures became available they were replaced by the compound engine.

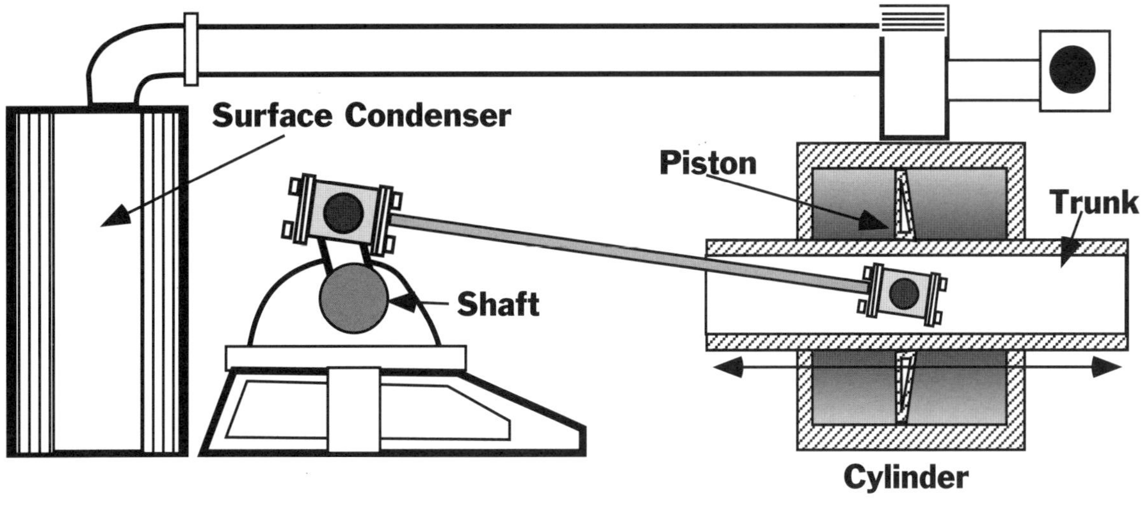

The trunk engine.
To save space a large hollow trunk was substituted for the piston rod. The connecting rod was connected to a journal or gudgeon in the center of the piston. The connecting rod was from 2.5 to 3 times the length of the stroke. They worked well with low pressure but under high pressure the trunks leaked. The large trunks created problems with the stuffing boxes. Weight of the piston would wear the trunks.
The diagram only shows part of the engine.

The steam tug Master. She is typical of the tugs that worked the B.C. coast and Puget Sound. She is an oil burner. The coal burners would look the same except for the bunker hatches. When she was working she belonged to Marpole Towing. She was engaged in towing scows. The log towers might have a bit less freeboard aft.
An oil burning tug of this size would have a 7 man crew. If burning coal she would have one or more firemen. They were a very quiet engine compared to a modern diesel. She is preserved by the SS Master Society, Vancouver B.C.

became the main engine used on nearly all screw-driven vessels. The correct name was vertical inverted direct-acting condensing engines. They were called inverted because of the older paddle wheel engines and some screw engines with the cylinder under the shaft. Having the cylinders over the shaft was simpler and more efficient.

It consisted of one or more cylinders mounted on columns over the crank shaft. The piston rod was connected to the crank shaft with a connecting rod and crosshead. Valves located on the cylinder were operated by valve rods. All the expansion of the steam was done in one cylinder. There could be more than one of these engines placed on one shaft. Each engine had its own condenser and air pump. A fly-wheel was used on some of them to make them run smoother.

It took some time before compound engines could be perfected. They were not considered very reliable until the 1870's. They ran faster than non-compounding engines, creating bearing problems. However this was better for propeller-driven vessels. They also used far less fuel. They became known as the reciprocating engine. The first ones worked at about 60 psi, the minimum pressure at which they could operate. Later compounds would use pressures of 80 to 100 psi. Most of them had the cranks set at 90 degrees to each other.

The compound engine was followed by the triple expansion engine. They were not really considered reliable until the 1880's. They needed a minimum of 120 psi to operate. The steam was exhausted from the HP cylinder to the intermediate (IP) cylinder and then to the LP cylinder. When steam is expanded to a lower pressure its volume increases. As the LP cylinder could get excessively large on big engines, the engines frequently had two LP cylinders. In later years, on some large ocean-going ships, a turbine was installed either alongside the main engine on its own shaft, or behind the engine to further use the available steam pressure before it went to the condenser.

A triple expansion engine could operate with around 200 psi with a condenser vacuum of 26 inches of mercury. The cranks were placed at 120 degrees to give a more uniform torque to the shaft. There were quadruple engines built as well to operate with a higher pressure and smoother power. It seems that most ships operated with 180 psi or less prior to WWII.

There were engines designed with cylinders on top of each other rather than the normal side-by-side configuration. One of these was the compound steeple engine. They were installed in some small tugs and are used in launches today. There were major problems with getting the engines on top center and keeping the valves synchronized. There was not much of a power gain with these engines either. Remember you cannot tell the power of an engine or whether it is a compound or triple by counting the cylinders.

These reciprocating engines were one of the mainstays of marine power for many years on vessels big and small. On the West Coast they replaced all the other types of steam engines except the horizontal engine used on stern wheelers.

While the principle was the same for all reciprocating engines, there were many designs. If you look at photographs taken before the turn of the century you will probably see a lot more large steam pipes running between cylinders than you would on a newer engine. This must have caused heat loss on the earlier designs.

During WWII many of them were built with the same or similar design for such things as Corvettes or Liberty Ships. They were often several

The engine room of the Master. As with most engine rooms on tugs there is no extra space to move around in. The controls on this boat are up on the grating at deck level but on a number of tugs they were in the engine room so the engineer could fire and oil the engine between bells (signals from the wheel house).

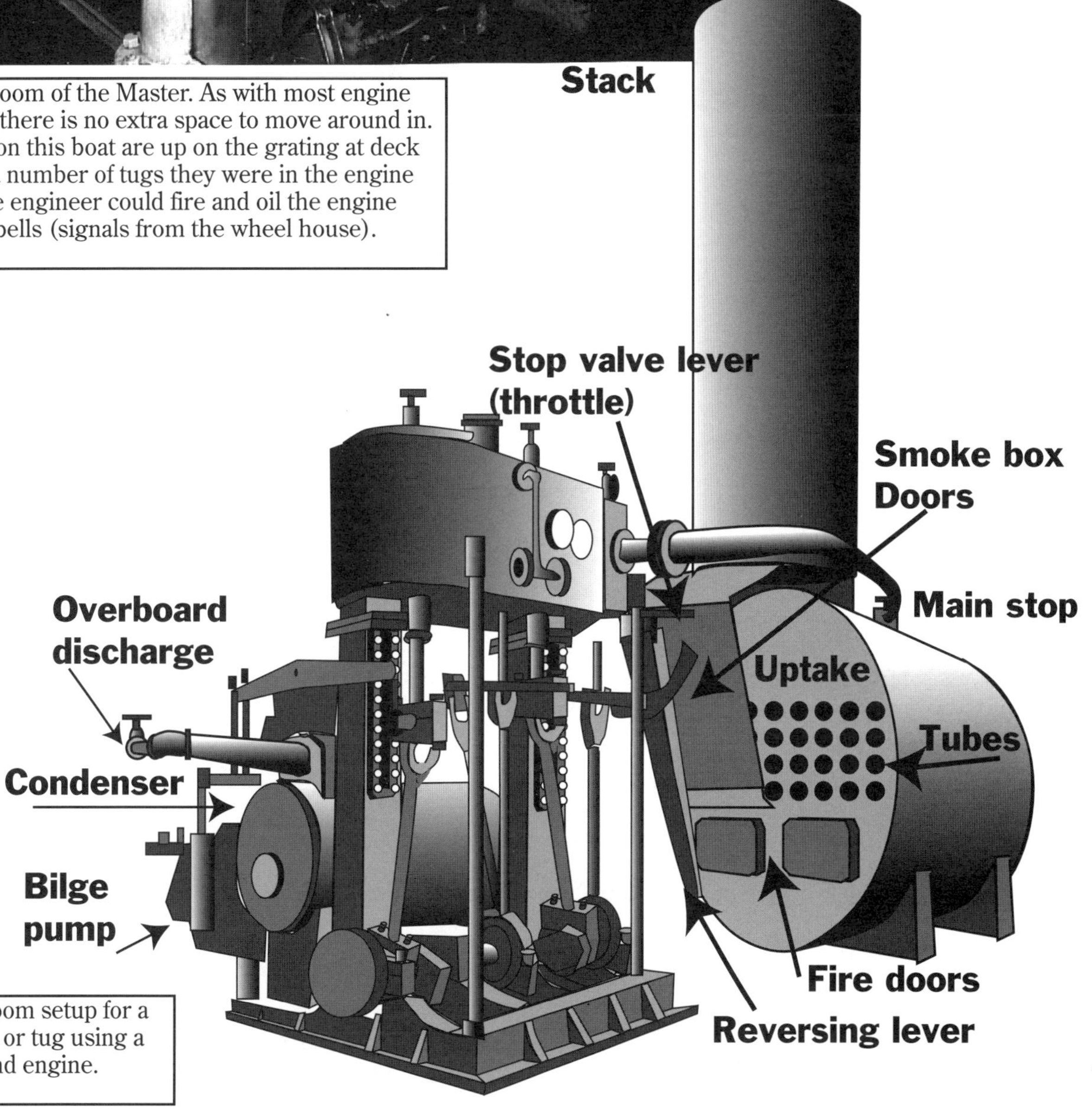

The engine room setup for a small coaster or tug using a compound engine.

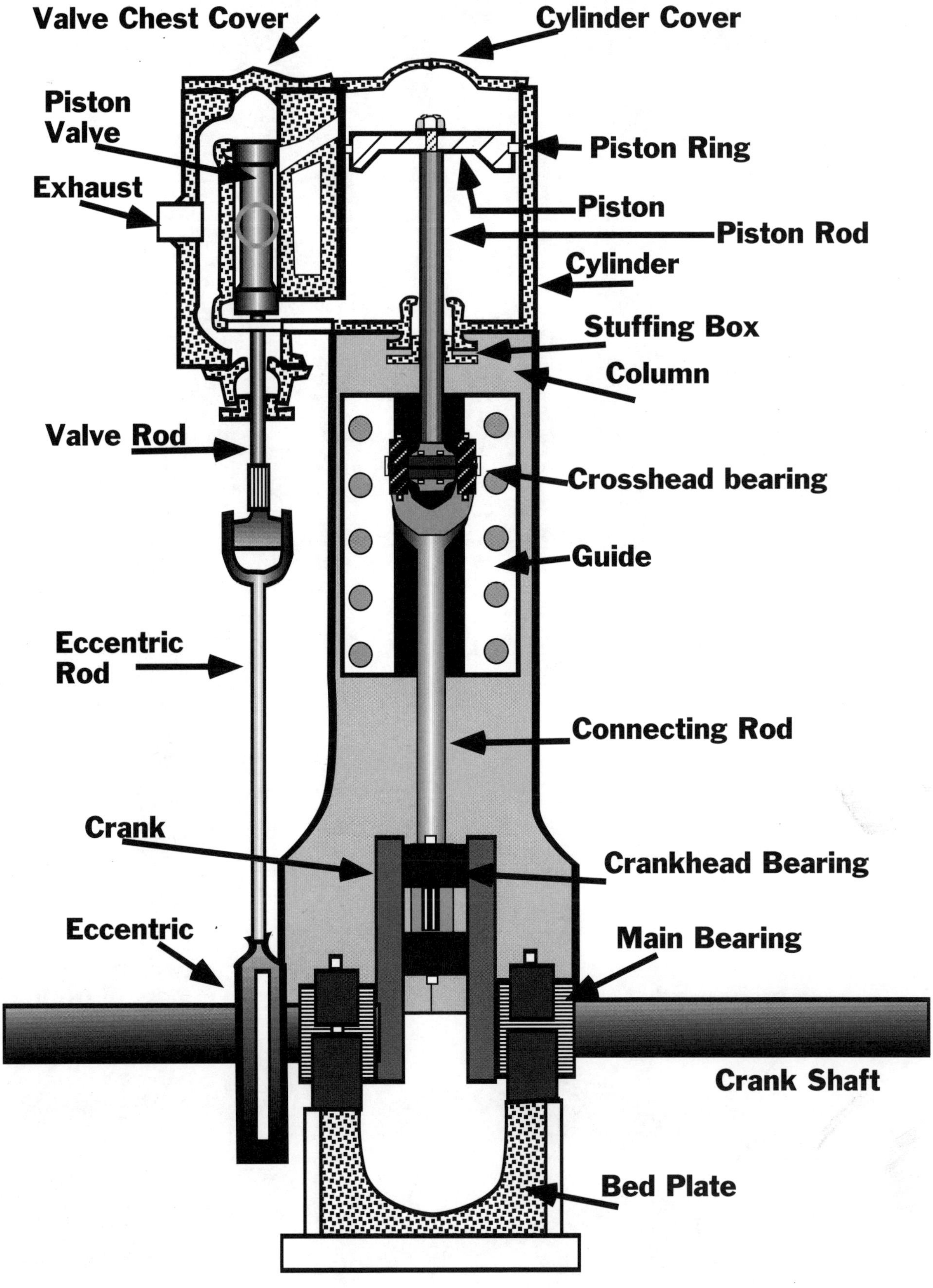

The names of the main parts of a reciprocating engine.

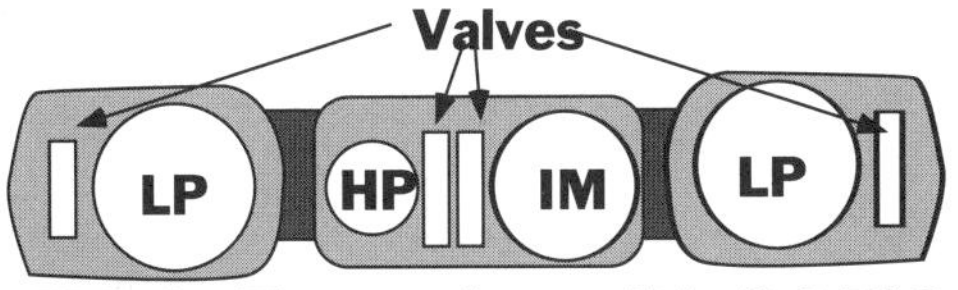

Triple Expansion with 2 LP'S

Compound

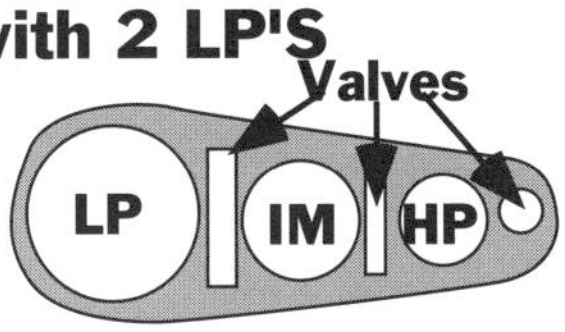

Triple Expansion

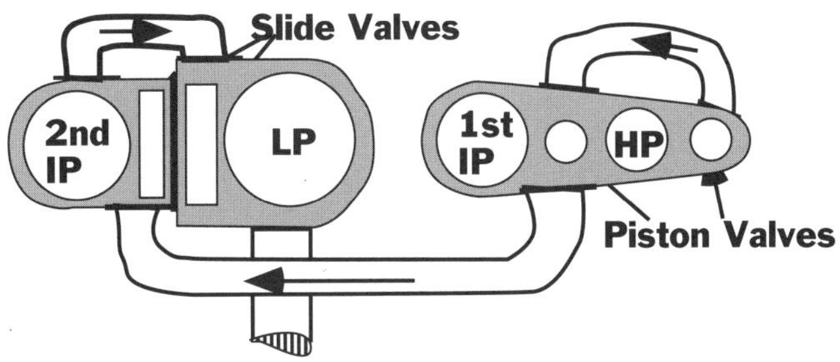

Quadruple Expansion Engine

> Engineers have built engines in many ways. Some of these are shown in the illustrations. Around the turn of the century cylinders were often stacked on top of one another. Marine engineers have told us this was not too efficient and made it difficult to adjust the valve timing. We have not seen examples of this in later engines. The triple expansion engine with 2 LPs was used on the Coast. Smaller vessels used compound or triple expansion engines.

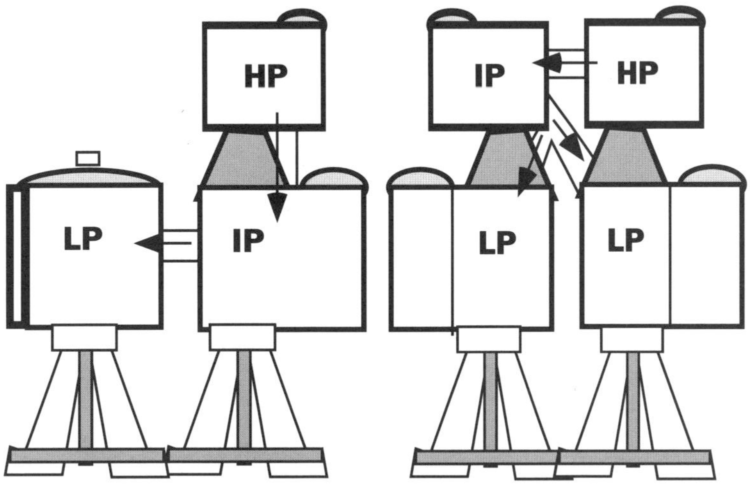

Triple Expansion

Compound

decks high when used in deep water ships. For example 19 ft. (5.8 m.) high for a Liberty Ship. They were heavy, 23,700 pounds (10750kg.) for a 2500 IHP triple expansion with a 4 ft. (1.2 m.) stroke. Smaller versions could be used in a tug without difficulty. They were a relatively slow turning engine (60 to 160 rpm) which made them good towboat engines. A tug needs a big slow turning wheel (propeller) to reduce cavitation when towing a heavy load. Today a Kort Nozzle is used to reduce cavitation when towing heavy loads. The steam engine of this type made it relatively easy to manoeuver a tug.

A variety of reciprocating engines, horizontal and vertical, were used for powering winches and other deck and engine room equipment. As reciprocating means back and forth motion a lot of small engines came under this category.

The engine was mounted on a bedplate. The base of the engine was called a soleplate. It was made of box-shaped parts bolted together. The main bearings for the crankshaft were bolted to the center. The columns were bolted to the outer part of the soleplate. The columns were of assorted design but in most cases there was at least one solid column for each piston to hold the guides. However some engines were designed with cylinders mounted on top of each other using one column and one connecting rod. In older and smaller engines the condenser was part of the back columns but on bigger engines, the condenser was a separate part mounted on brackets. All these parts were carefully lined up before the cylinders were mounted on the columns.

The columns could be double, port and starboard, or single. If the usual right handed propeller was in use and only one column was used per cylinder, the column was on the port side to take the stress on the guide when going ahead.

The cylinders were made of cast iron or steel. Often liners were fitted for the pistons and the valves to work in. A cover was fitted at the top of each cylinder and valve. Again, the design varied from engine to engine. Some had a manhole cover. Safety escape valves were mounted on the top and bottom of the IM & LP cylinders to let water out or relieve excess pressure. They had stuffing boxes for the rods at the bottom of the cylinder and valve casings. The piston was usually an iron or steel casting. Pistons had a number of piston rings installed. This often included a "junk ring" to make changing rings easier.

Each cylinder needed to have a slide or piston valve enclosed in a steam box beside it. Cylinders and valve casings had ports cut into them. The location and size of ports depended on the engine design but they were designed so they would not cause a problem for the piston rings. There were many types of valves used to let the steam in and out of the cylinders as required. The duties of a valve were, to admit steam to the cylinder, cut off the supply of steam, open the port to exhaust and to close the exhaust port. Most of them were slide

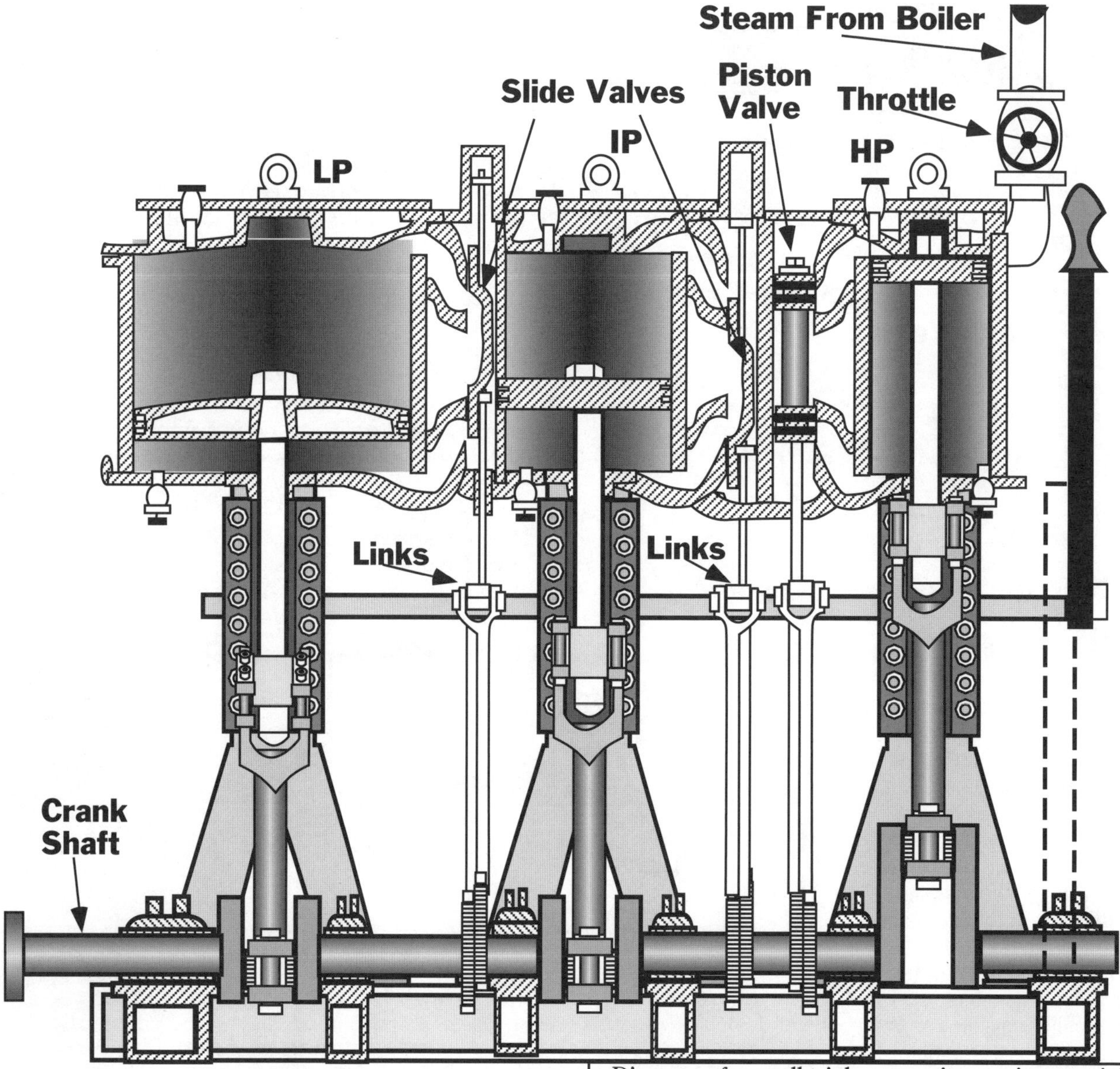

Diagram of a small triple expansion engine as might be found on a number of vessels under 200ft. (61m.). Pumps not shown.The thrust bearing is connected to the LP end.

or piston valves. There were many types of these valves. The HP valve was often a piston valve while the other valves might have been slide valves. On large engines, cam operated poppet valves could be used on the HP cylinder. In later deep sea ships balanced valves were used to allow for the high pressure.

When this part was machined the cylinders were usually bolted together. The cylinders were mounted on the columns. The cylinders were usually covered with felt to reduce heat radiation. The felt was covered with wood or sheet metal. This is why the cylinders look like they are all one piece to the casual observer.

The pistons were connected to the crankshaft by piston rods and connecting rods. The piston rod was worked in a guide on the column to keep the piston lined up. The piston rod was joined to the connecting rod by a crosshead bearing.

The bearings needed continuous lubrication. There were a number of methods used to accomplish this. This could be done with individual cups or a lubricator.

Model of a triple expanson engine at the San Francisco Maritime Museum. San Francisco Maritime Historical Park.

Cylinder Heads

Relief valve S.S. Master. On the bottom of the cylinder.
Relief valves are located on the top and bottom of the cylinder

The cylinder heads from 3 triple expansion engines. The top one is the low pressure and intermediate pressure cylinder of a Liberty Ship. The valve covers are located between the cylinders.The cylinders are insulated to reduce heat loss so they look like they are all one.
The middle one is the cylinder heads of the tug Master. The parts are much smaller but the principle is the same. The relief valves are used if water gets in the cylinder or steam box or if the pressure becomes too high.
The bottom photograph is from a launch engine. Relief valves are located on the top of most of the cylinders.

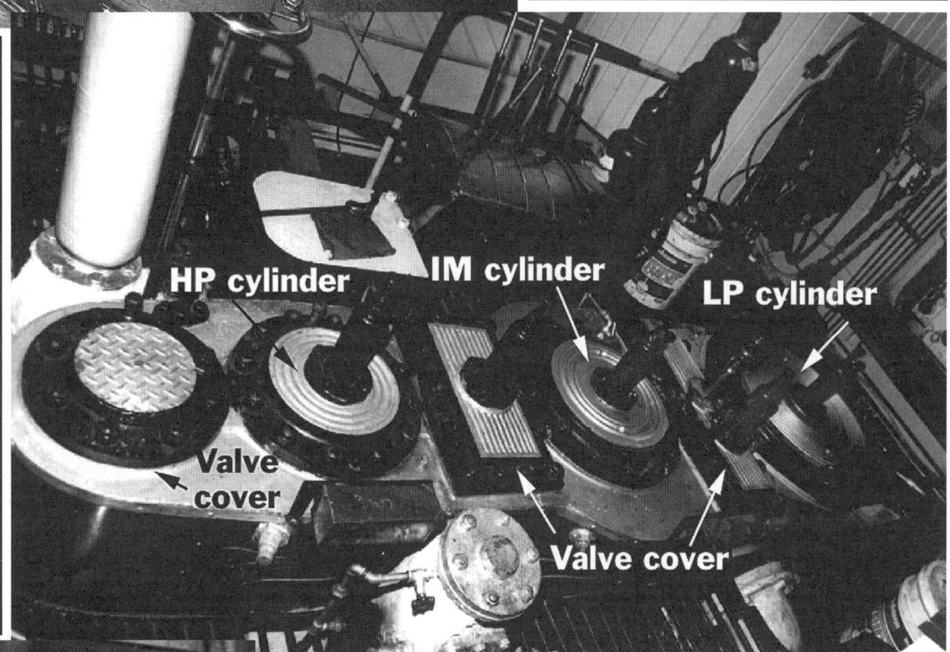

On some vessels the bearings had to be oiled by hand. There were brass boxes fitted with a syphon pipe leading to the bearings. The box had what is known as worsted.in them. Small tubes would feed the pipe with oil by capillary action. The oiler learned to get the rhythm of the machine as he worked with his oil can keeping the oilers filled. The rods had to be swabbed with a heavy lubricant. The eccentrics where often lubricated with the use of a pan. Tallow and later grease in cups with screw down tops would be used on some surfaces. On large ships, a form of power lubrication was sometimes used. It was safer and more reliable.

The internal parts of the cylinders and valves must be lubricated as well. Again this was accomplished in several ways. On smaller and older engines the oil from the rods mixed with the moisture from the steam was sufficient. Not much oil got back to the boilers this way.

When the engine speed and the pressures increased, lubricants had to be introduced to the cylinders. With dry saturated steam, lubrication was a must. This was done by forcing oil into the steam pipe or valve chests with a pump or using some of the steam to put oil in the main steam pipe.

The crankshaft was connected to a thrust block at the back of the engine by a shaft. Shafting connected to the thrust bearing went aft through the stern bearing and stuffing box to the propeller or "wheel" as it was referred to. This shafting was in sections held together by couplings and supported

The bottom end. The connecting rod moves the crank. Note the tubes running down the side of the connecting rod for lubricating the crankhead bearing. They run up to brass boxes with oil in them. The engineer or oiler felt the bearing with his hand to see if it was getting hot. Normally there would be an oil box on the main bearing. It is missing in this photograph.

Oil is placed in the brass box (oil cup) with an oil can. These oil cups are often worsted. That is there is a series of small tubes connected to a siphon. This lets the oil drip slowly on the bearing it is lubricating.

by a series of bearings. These bearings had to be lubricated and the shafts had to be kept lined up.

Engine Operation

The steam from the boiler and the main stop valve, went directly to the engine through the main steam pipe and the intermediate stop valve. There may have been a separator to take out any water installed in the steam pipe. The amount of steam entering the engine was controlled by the engine stop valve (throttle). Steam was introduced through

Crossheads where the piston rod connects to the piston rod. The bearings are lubricated with oilers. Some of the big engines had large lubricators.

Columns support the cylinders and valves. The piston rod guides are located on them. They take a lot of pressure on a high powered engine.
On larger vessels a double column is usually used.

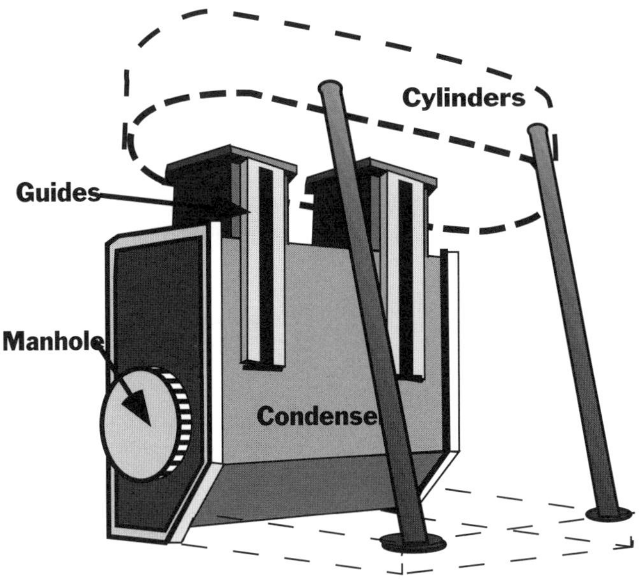

Combination Condenser and column

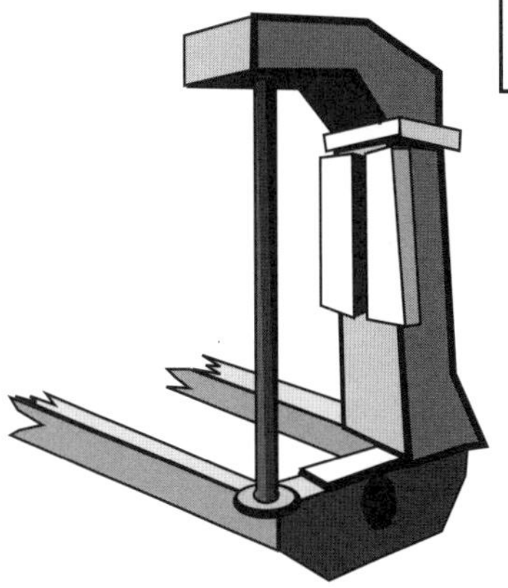

Single Type column with open front

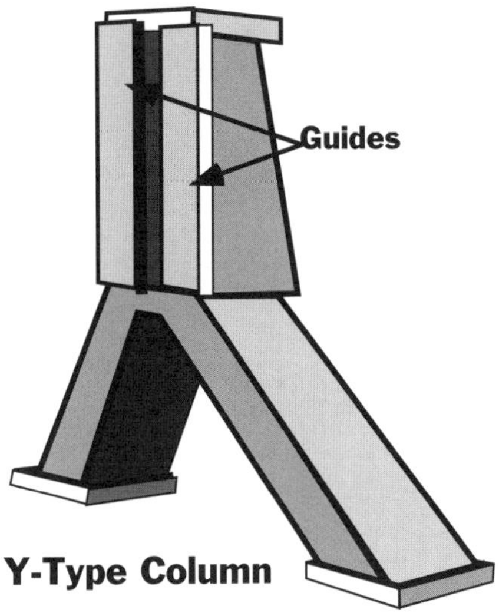

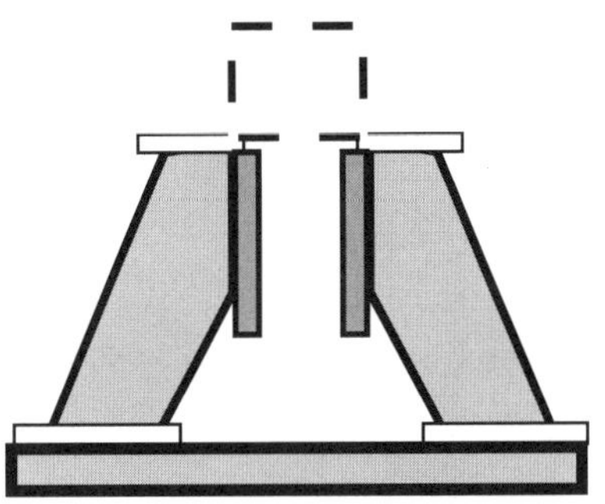

Double Type Column

Y-Type Column

one port while it was exhausted through a port at the other end of the cylinder. The steam was cut off before the piston reached the other end of the cylinder to let the expanding steam do the work. This was regulated by valves located in steam boxes beside the cylinder.

These valves were controlled by valve rods connected to the eccentric bearings on the shaft. There were two rods and bearings for each slide valve, an ahead rod and an astern rod. This controlled which way the engine ran. The required rod was put in place by moving the "link block." The link system was moved by a control system such as a Johnson bar in small ships, or by power (a reversing engine) in larger ships. When manoeuvering, the engine was brought to a stop, the eccentrics changed and the engine restarted in the opposite direction. The engineer could get the engine stuck on "top center" as with other engines. The engine could not go ahead or astern until the crank was moved often by hand. By-pass valves to the other cylinders were fitted on large engines. This problem caused the odd crash.

Some pumps were usually operated by the main engine. These were the air pump, bilge pump, circulating and feed pump. One or more of each of these pumps operated simultaneously. They were normally operated by the IP cylinder. Two plates mounted to bearings on the IP column worked a crosshead up and down. The crosshead operated the pumps.

After the steam left the engine it passed to the condenser. As with other types of steam engines the condenser on a reciprocating engine was as important as any other part of the engine. Steam and water were pulled into it by a vacuum created by an air pump. The steam contacted tubes filled with circulating sea water. This condensed the steam to water were it was drawn from the bottom of the condenser to the hot well. It would be injected into the boiler as needed. Sea water entered the condenser at the bottom and was discharged from the top after going through the tubes. The end plate could be removed for maintenance. There were several condenser designs but probably the most common was one that looked like a big cylinder. The cooling surface was around 1sq. ft. (.09sq. m.) per IHP (indicated horse-power) with a vacuum of 25 in. of mercury. This required a number of tubes.

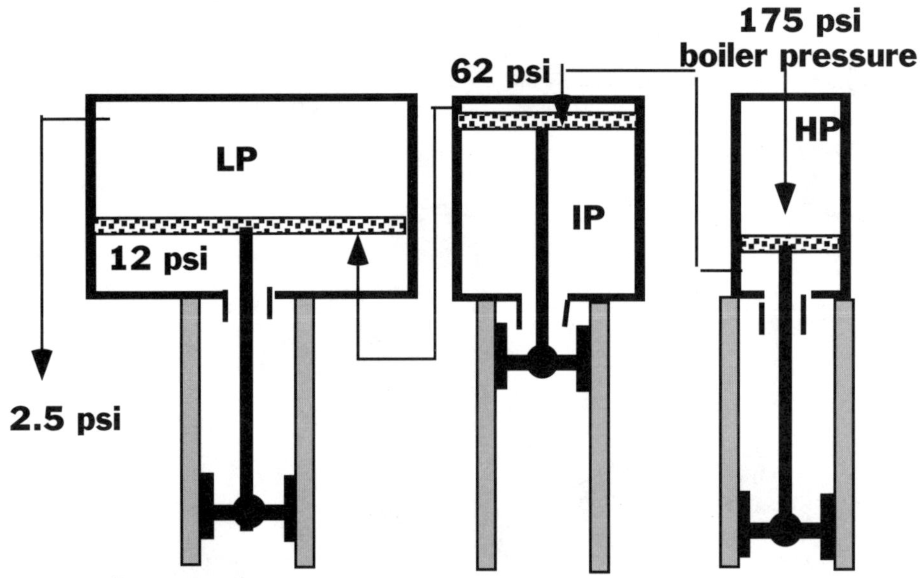

The steam flow in a triple expansion engine. The pressure is reduced in each stage as the steam expands. The valves control the steam entering and exhausting the cylinders. The pressures are approximate. We have not shown the back pressures involved. The figures are from "verbal Notes and Sketches".

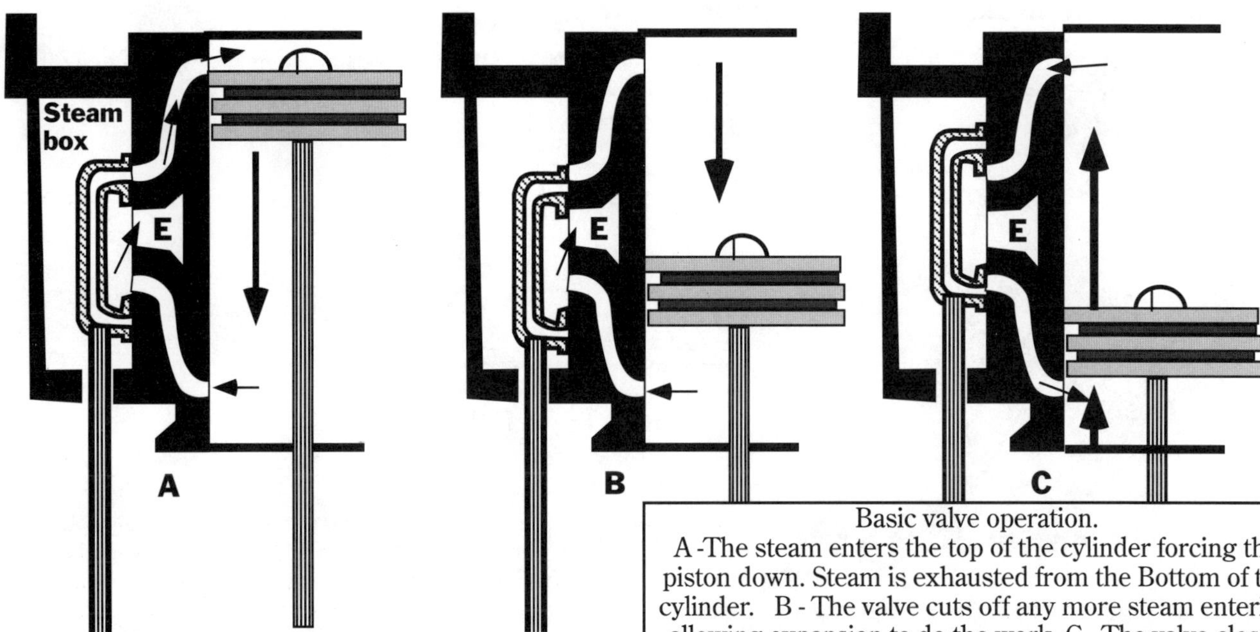

Basic valve operation.
A -The steam enters the top of the cylinder forcing the piston down. Steam is exhausted from the Bottom of the cylinder. B - The valve cuts off any more steam entering allowing expansion to do the work. C - The valve closes the exhaust port at the bottom just before the piston reaches the bottom so there is a slight steam cushion. The valve moves a little further allowing steam into the bottom of the cylinder and steam out at the top. The amount the valve is open to steam when the piston is at the end of the stroke is the lead. Some engineers allow more lead on the bottom of the cylinder for a greater cushion. Valve timing is very critical and requires an experienced engineer to do it. The valves are moved by small eccentric bearings on the crank shaft through valve rods and eccentric rods. The valves determine which way the engine is running, ahead or astern.

The maximum vacuum available was a necessity for engine efficiency, maximum power and fuel saving. The amount of vacuum would rise and fall with the barometer. Sea water temperature also affected the vacuum. An increase of 10ºF in the sea water being pumped through the condenser would reduce the vacuum about .75in. By checking the barometer and the vacuum gage the engineer could, after doing some calculations, determine how much atmospheric pressure had been destroyed in the condenser.

The water went from the condenser to the hot well before being returned to the boiler. The water was kept warm to avoid putting cold water in the boiler. The water was filtered at this stage to remove oil from the engine. Some steam was used for running other equipment but when possible, the steam was returned to the condenser. If the engine was not running such as the vessel was at anchor, the steam used by auxiliary equipment was lost unless the vessel had an auxiliary condenser. This wasted valuable fresh water. Most deep sea ships could "make water" with an evaporator but as a rule, coastal vessels had to take on more fresh water from ashore.

On the old steam tugs, the engineer would shut down the dynamo (generator) as soon as the boat was tied up to prevent water loss. As they usually did not have any batteries, this meant instant darkness if the oil lamps were not lit. This often happened before the deck crew had lit the coal-oil

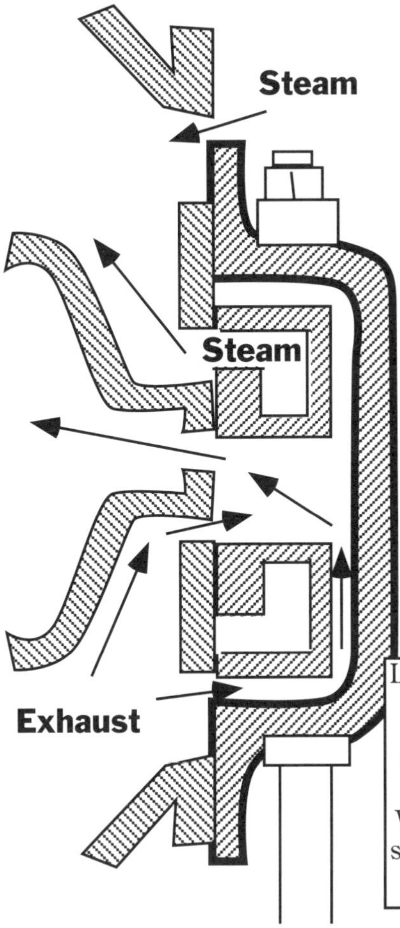

Left - A double-ported valve. The valve has less movement than the D valve and works better with a faster engine. There are many slide valve designs. When using superheated steam steel valves were needed instead of the usual iron ones.

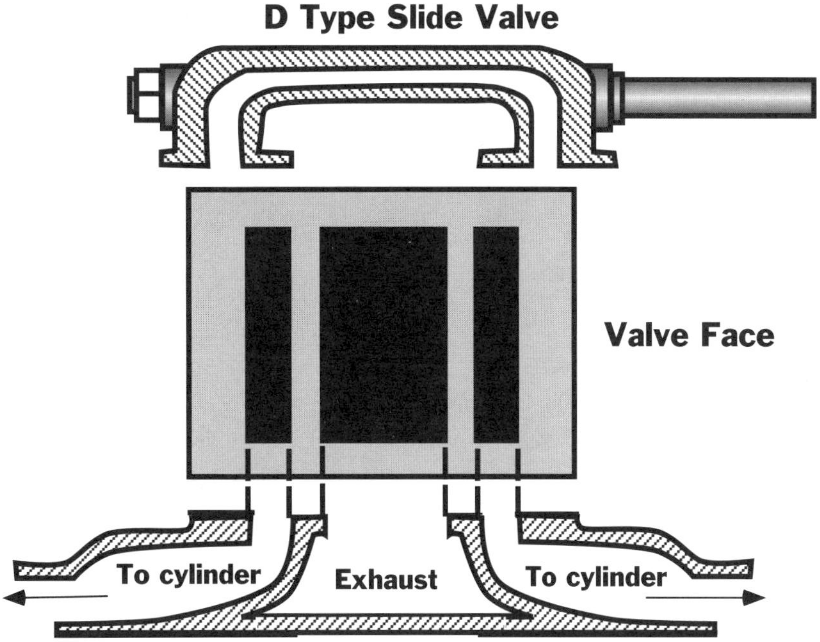

Above - The valve face for a D type slide valve.

Valve face on a double- ported valve

The valve face of an inclined engine. San Francisco Maritime Museum.

The eccentric bearings and straps on the crank shaft. One is for ahead and the other is for astern. One of the reversing linkages is used to connect the correct one to the valve rod. A Stephenson link is used in this case.

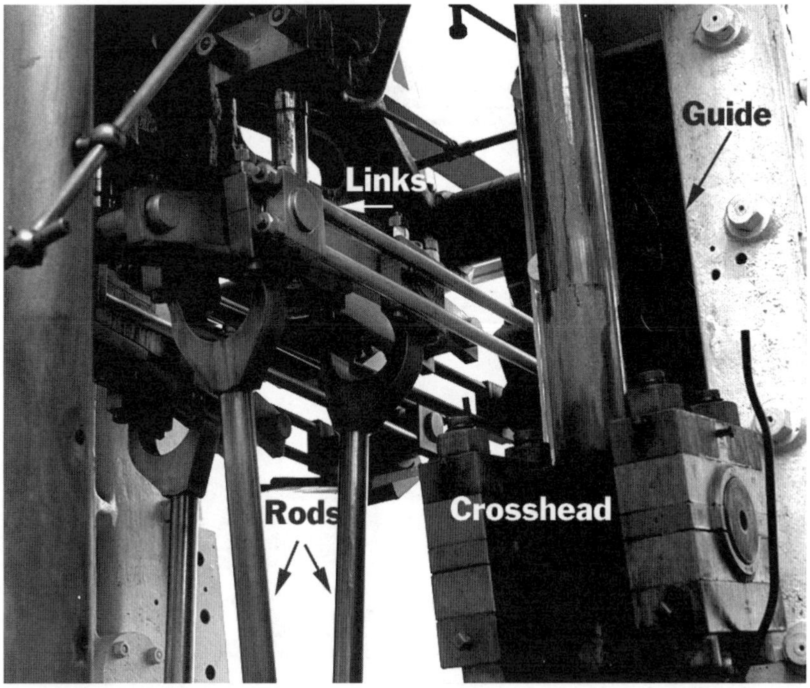

The linkage on a triple expansion. Toronto Maritime Museum.

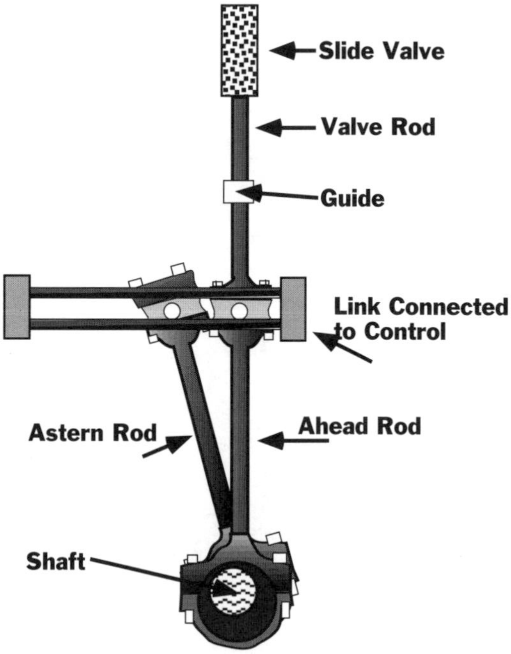

The Stephenson link. The link is moved by the control mechanism to engage either the ahead eccentric or the astern eccentric with the valve rod. When the engine is stopped the engineer moves the link to the ahead or astern position. When steam is applied to the engine again the valves will start the engine turning in the right direction. The Stephenson link in one of its many forms is the most common method of valve control but it is by no means the only one. There are several types of radical valve gear systems including the joy valve gear which will allow reversal and a variable cut-off.

lights or got the deck secured. The odd bit of bad language could be heard as the seamen stumbled around in the dark trying to find a lantern to light. On some small passenger/freight boats not fitted with an auxiliary condenser, a separate circulating pump and air pump were used. They were operated by their own engines.

This way very little fresh water was lost. A large engine used a lot of water. A WWII destroyer steaming at full speed used 200 tons of water an hour (from 1941 RCN "Machinery Handbook").

On later ships there were very complicated systems in place for converting the heat from exhaust gases to steam for a variety of uses. The system used depended on the type of ship (tanker, bulk cargo, or general freight) owner and the flag.

To avoid engine racing which could be caused by a broken shaft, propeller loss, or the vessel working in a sea, a governor was fitted on some engines. There were a number of makes available. Most of them had to be adjusted for the existing sea conditions. In very heavy seas the engineer still had to standby the throttle to keep the engine from racing when the propeller lifted out of the water.

As with other types of engines, the engine room staff was kept busy at all times. The engineer had to keep ahead of repairs and adjustments as well as overseeing the crew. When docking he had to handle the engine. The oiler had to keep all the machinery lubricated and checked for any hot bearings. The wiper kept the engine room clean. Everybody listened for any changes in the sound made by the machinery. An unusual sound meant something could be wrong. On tugs and small coastal freighters the engineer would have to look after all these jobs as well as possibly firing.

The eccentric rods. On a Liberty Ship

Unaflow

Another type of engine that did operate on the coast was the "Unaflow." In theory it eliminated initial condensation the major problem with reciprocating engines. With the unaflow engine the steam entered the cylinders at the ends after going through heads that were warmed by steam jackets. When steam was expanded it exhausted through ports in the center of the cylinder. These ports were

Reversing Gear

Reversing the engine quickly is of prime importance on a vessel regardless of size. On small tugs and freighters this could be done with a reversing bar to move the links. On larger engines a reversing engine of some type was used.

The terminology used in relation to engines changes from time to time but the variety of names for the parts in reversing gear is large. The name can change in the same manual.

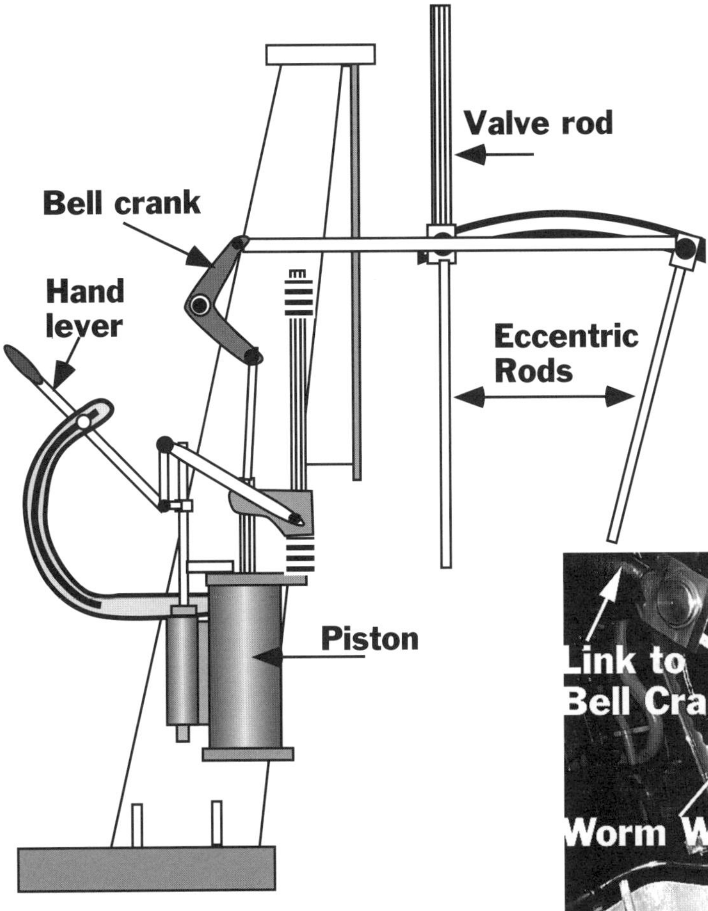

Type of reversing gear favored in the U.S. When the hand lever is moved it activates a piston or slide valve. This allows steam to move the piston to the other end of the cylinder. The piston rod moves the bell crank and the rock shaft (sometimes called weigh shaft or tumbling shaft). The various links connected to the bell cranks are shifted to the ahead or astern position.

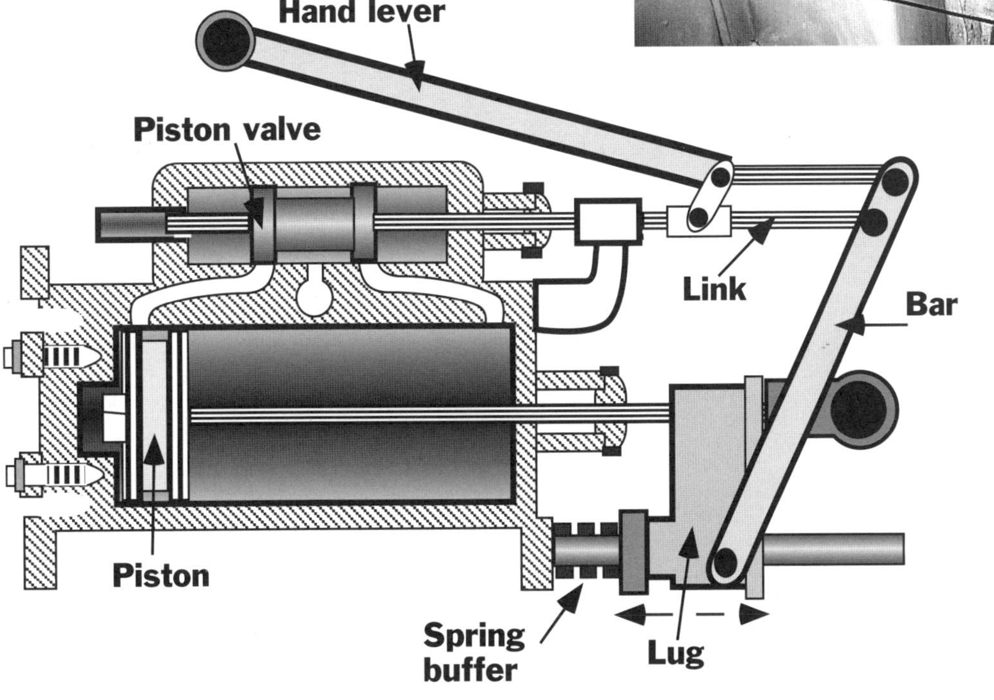

Above.
The type of reversing found on a lot of English ships and Liberty Ships. The reversing engine is controlled by a lever. It drives a worm shaft geared into a large worm wheel which is attached to a link to the bell crank. The fly wheel keeps the movement from being too violent. Hydraulic and steam assists were used on some reversing gear. They were much easier to adjust.

The controls on the Virginia V. These were typical of many tugs, small passenger ships, and freighters. The reversing bar (or Johnson bar) was pulled over by hand turning the rock shaft (or weigh shaft) when the engine was stopped. The bell crank (or reverse shaft lever) moved the links. The amount of steam was regulated by the throttle valve.

Other types of reversing gear

The Stephenson link became the most common method of controlling the valves but other systems were used. The Joy valve gear used levers connected to the connecting rod with no eccentrics. Crank valve gear used gears to control the valves. Marshall gear used one eccentric and a gear. Walschaert valve gear used one eccentric and levers.

The condenser on the Master. After the steam leaves the LP cylinder it goes to the condenser where it becomes warm water. The air pump removes water and air. The water goes to the hot well before going back to the boiler. A vacuum is maintained in the condenser at all times.

Pumps on the S.S. Master. They are connected to the engine by levers and a crosshead. They are the feed pump, the air pump and the circulating pump. The circulating pump supplies salt water to the condenser.

The large pipe taking the steam from the low pressure cylinder to the condenser on a Liberty Ship. Steam from other sources such as auxiliaries are also fed back into the condenser.

Sea cocks on the skin of the tug Master. They allow a flow of salt water to circulating pump and the prime for other pumps. They are usually gate valves that can be shut off when needed.

The Auxilary condenser on a Liberty ship. Steam can be sent to it when the main condenser is shut down.

uncovered by the piston at the end of the stroke eliminating exhaust valves on some engines. They had a long piston equal to half the stroke and half the width of the exhaust port. The steam flowed in one direction.

There were a number of these engines developed in different parts of the world. Some used compression for part of the stroke with the steam going to a condenser. Others were non-condensing and some had exhaust valves.

The steam valves of the double-beat poppet-type were quite complicated. Cams were used to activate them. Some engines used Corliss-type valves. The control of speed, stopping, starting and reversing were done with the cut-off valves. The steam valve was left open. Most of the engines operated in an enclosed space with forced lubrication as you might find on a diesel.

Some of them were used on horizontal engines for sternwheelers but as far as we can tell not on the West Coast. At least one coastal passenger ship had this type of engine. They did not seem to be a huge success. One engineer we know, said they were difficult to keep running.

Steeple Compound Engine

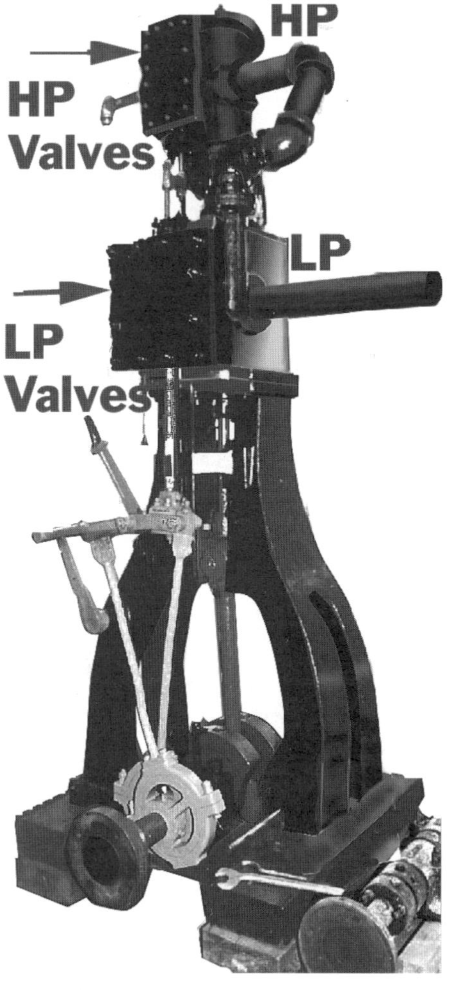

Steeple Compound Engine. This engine was used on a small tug the "Charlotte" on Lake Whatcom. This is an old engine and typical of the type used on small tugs before diesel engines. She had a jet condenser for use in fresh water. The tug had a watertube boiler. Whatcom Maritime Historical Society Bellingham Washington.

Steeple Reciprocating Engines

These engines were put into a number of small tugs to conserve space. This was important for small yarding tugs working in booming grounds.

Steeple compound engines should not be confused with the low pressure steeple engine used in the paddlewheelers. The steeple compound is a reciprocating engine that has one crank with the high pressure cylinder directly over top of the low pressure cylinder. The valves are located beside the appropriate cylinder. Both valves are operated by one valve rod. This made it difficult for engineers to time the valves. The distance of the top HP valve from the cylinder reduced the efficiency on some of the earlier ones.

With only one crank the engine could easily get stuck on top center. The engineer would then have to turn the engine with a bar. The single crank did not run as smoothly as one with two or more cranks. This was overcome in several ways including the use of a flywheel and weighted webs. Some types such as the Wells balanced compound used two cranks and weights to overcome these problems.

Some of these engines were built in Victoria. They were not an efficient engine and were not built for too many years except as small launch engines.

Diagram shows how one piston rod is used and how the two valves are controlled by connected valve rods. The steam exhausts from the HP exhaust port to the LP steam box.

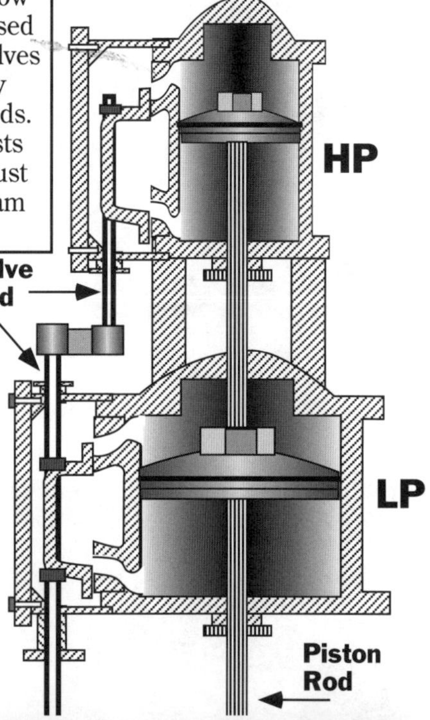

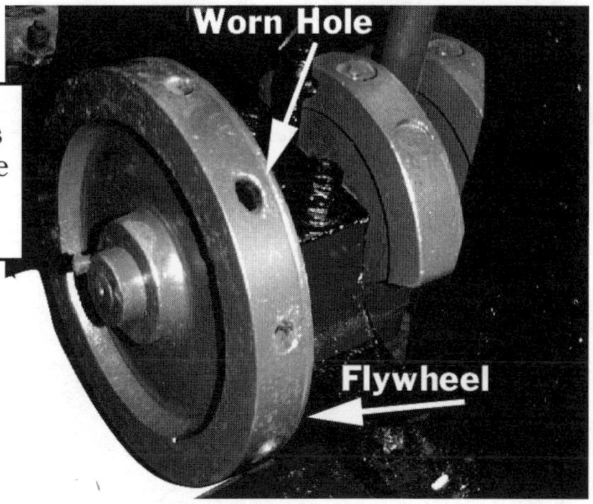

The flywheel and weighted webs on the crank make the engine run smoother. Steeple compound engines have a habit of sticking on top center. The hole for the bar was badly worn on this old engine.

The steeple compound used in a modern launch in Victoria B.C.

Turbines

Steam turbines have been around for a long time in one form or another. They replaced other types of engines in faster vessels and in later years of steam propulsion were used with reciprocating engines. Turbines are used on nuclear powered submarines. Some very large ships such as tankers are still powered with turbines. They are used for other purposes on board a number of ships. They have totally replaced the reciprocating engine on modern steam ships.

Classification of Turbines

1 Impulse
2 Reaction

1 Simple - one stage
2 Compound - expanding in stages

1 Single stage
2 Multi stage - step down pressure
1 Raviac - inward and outward flow
2 Axial parallel mix flow

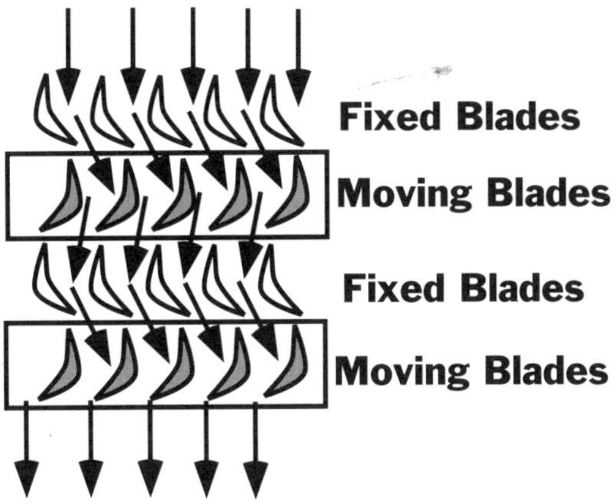

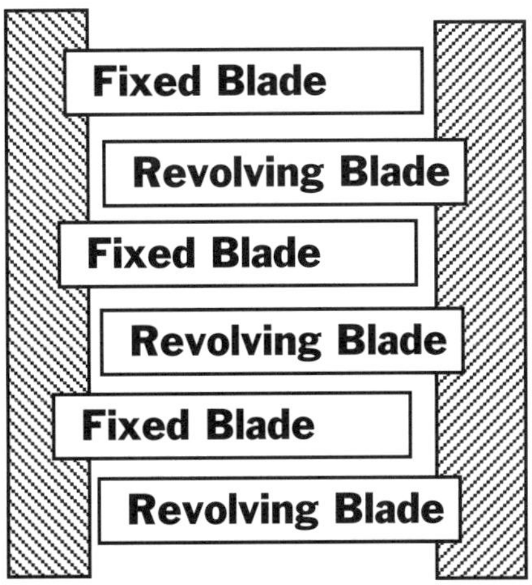

The basic theory of a turbine is steam expanding past fixed and rotating blades makes the rotor turn at high speeds.
The vanes operate like nozzles changing the direction of the steam. The centrifugal force produced by changing the direction of a jet of steam escaping from a nozzle at high velocity moves the rotor and shaft. This is accomplished by placing the steam nozzle so it will impinge on numerous curved vanes attached to a wheel (rotor) that is free to revolve.

The rotary engine had been invented in the 18th century but it was not practical until 1884 when Sir Charles Parsons invented a turbine to operate an electrical generator. The first one to power a vessel was in HMS Turbinia in 1897. This was a direct-drive engine operating at a relatively slow speed. The Curtis turbine was invented in the United States.

It took some time even after they were proven for engineers to accept them. To quote from the Manual of Marine Engineering 1904, "Notwithstanding all that is said of them by the patentee and his supporters may not be accepted without question by engineers generally, it must be admitted by all that these motors possess some distinct advantages...." The main problem was their high speed and they could not be reversed.

By 1905 the R.N. had put them in a lot of their new ships. Fast passenger ships quickly adopted them. After WWII most steam merchant ships were turbine powered. There were not as many used in coastal service.

They are very sophisticated pieces of equipment so we will only touch on the basic principles.

Types of turbines:

(1) Impulse turbine.
(2) Axial flow reaction turbine (Parsons).
(3) Axial flow impulse reaction.
(4) Radial flow reaction.

A turbine is a form of heat engine. In its simplest form it consists of a rotor with blades revolving in a casing. The casing has nozzles or stationary blades to direct the expanding steam to the rotor blades.

If a jet of steam is directed on a vane, placed at such an angle that the direction of the flow is changed, then a force or impulse will be exerted on this vane. In a turbine, suitable vanes or blades are arranged on a drum or wheel which is free to revolve inside a casing. A flow of steam is arranged to pass through these blades in such a manner that a force is exerted on the blades, thus causing the drum or wheel to rotate. In addition to the blades on the drum, there are generally rows of blades on the casing to direct the flow of steam.

To help understand some of the theory, we have paraphrased parts of the data found in the machinery handbook (1941.) If there is a hole in a vessel containing high pressure steam, it will produce a high velocity jet of steam. The velocity of this jet can be increased if the steam escapes through a specially shaped aperture, known as a nozzle. The velocity will depend on the difference in the pressure of the steam in the vessel and the space into which it is escaping. If steam is expanded from 250 psi. to condenser pressure through a suitable shaped nozzle, it will attain a velocity of over 4000 feet per second. This is about twice the velocity at which a bullet leaves a rifle.

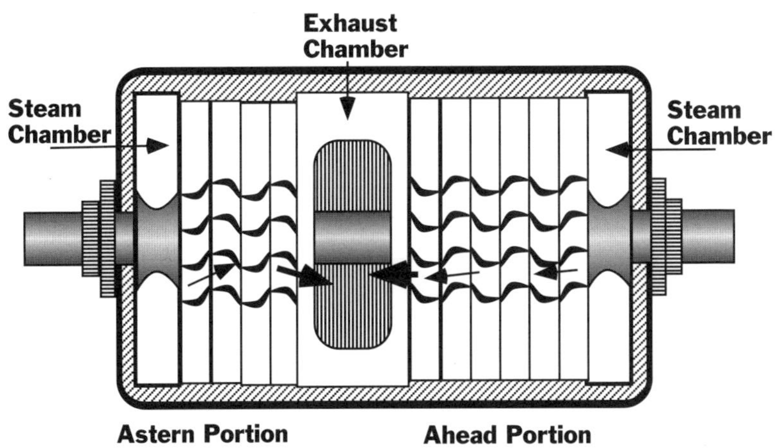

Diagram of one of the first Parsons' turbines.

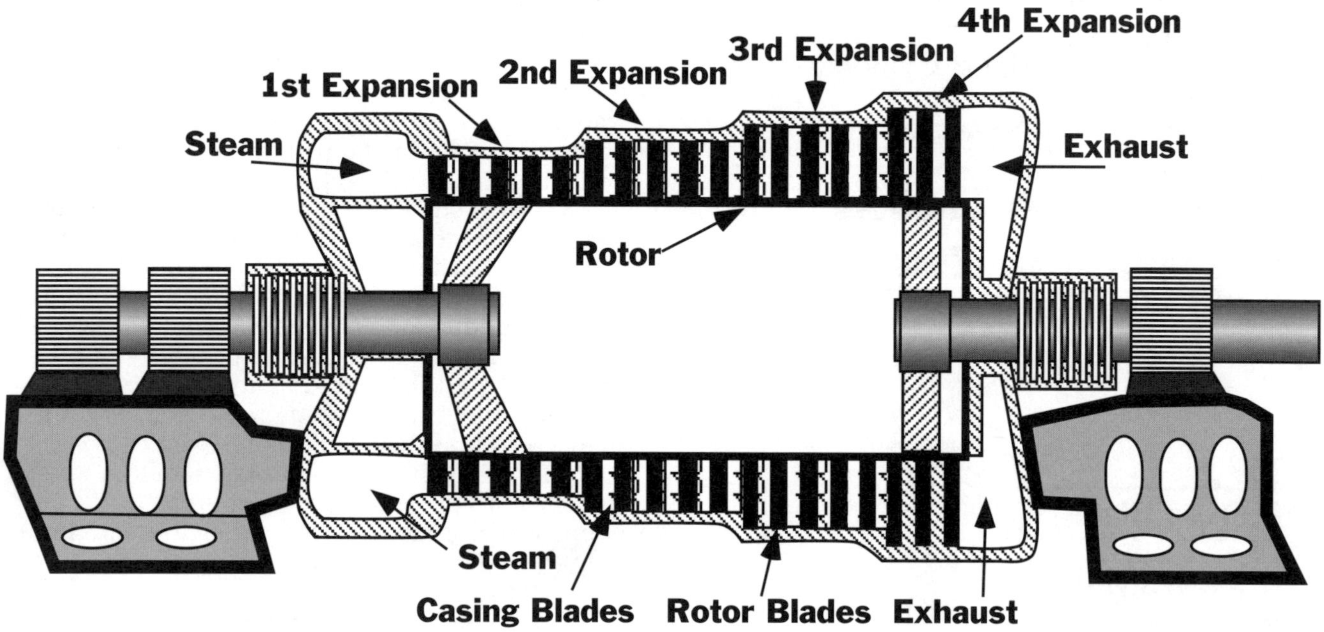

A very simplified diagram of a Parsons' HP turbine.

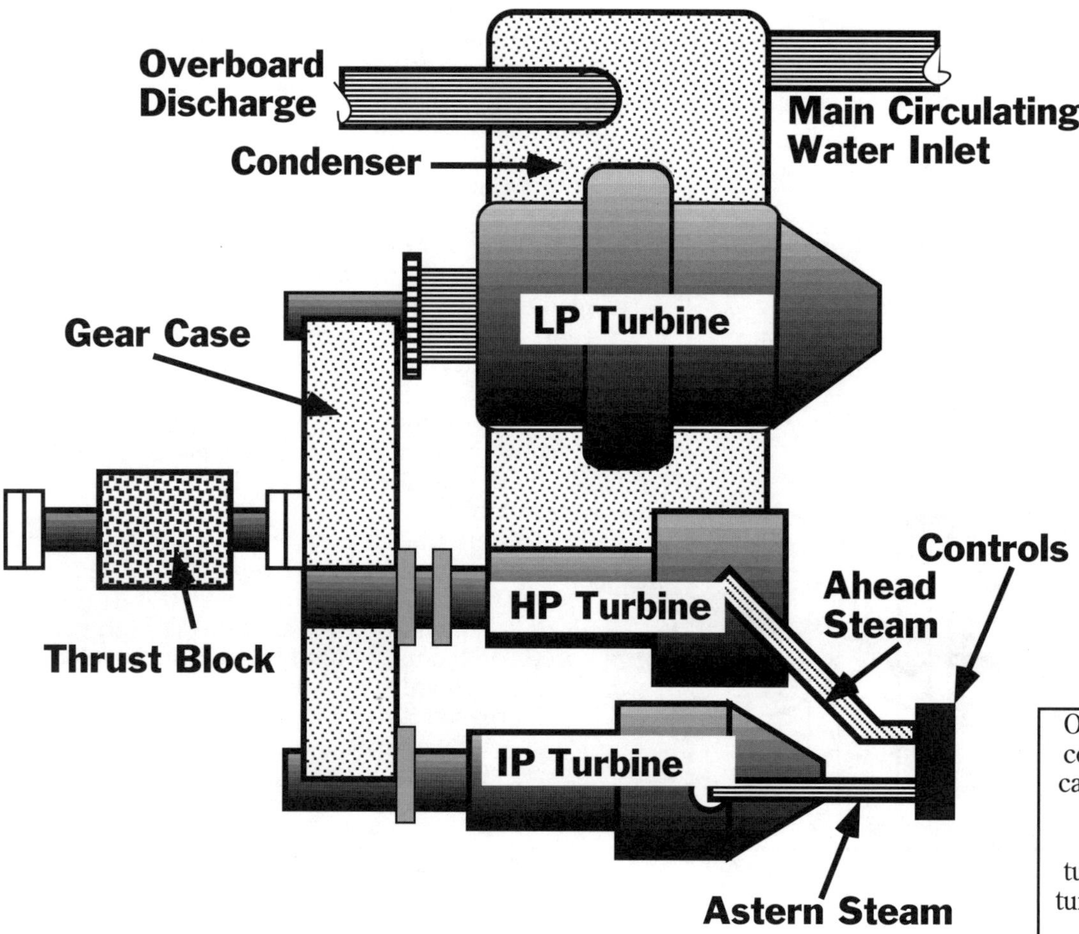

One of the many ways a combination of turbines can be used in an engine room. Often one HP turbine and two LP turbines are used. Each turbine can be on its own shaft.

A number of things are unique to this type of engine. A separate turbine is used for going astern as they cannot be reversed. Some models have an astern section built into them. This reduces the astern power considerably. The condenser vacuum is extremely important. Due to the high speed of rotation, most turbines are used with gearing to the propeller shafts so the propeller shaft RPM will not be too high.

There are assorted types of turbines such as impulse and reaction turbines but the basic theory is the same. We will show just one of the simpler turbines. If research is being done on a particular ship, it probably will require consulting the manuals on the type of machinery involved.

Low pressure turbines were installed with reciprocating engines on some ships. This made better use of the steam exhausted from the LP cylinder by using it in the turbine before sending it to the condenser. When in later years better materials were used in reciprocating engines superheated steam could be used without a lubrication problem in the cylinders. The two types of engines could be used on the same shaft if an oil filled coupling was used.

On some ships a high pressure turbine was used with two low pressure turbines each connected to its own propeller shaft. Some turbo-electric ships used the turbines to generate electric power for the electric motors on the propeller shafts. All pumps and electrical generators could be operated with their own turbines.

Some advantages: Ability to produce high horse power, lack of vibration, superheated steam could be used effectively as they required no internal lubrication, no internal initial condensation and the center of gravity was low. Some disadvantages: separate turbine for going astern, difficult to operate at slow speeds and if there was trouble with any of the blades the engine would have to be shut down.

Other Types Of Turbines

There have been a number of makes and designs used at sea over the years. Some of the other popular makes in use before 1960 were the Westinghouse, Curtis and Rateau turbines.

Curtis turbines has its roots dating back to the 19th century. They were an impulse-type high speed geared turbine. Steam expanded in fixed nozzles which impulsed on the moving rotor blades. A high steam pressure could be used. The pressure was dropped from the nozzles at high velocity through two sets of rotor blades. This was repeated in up to four stages.

Rateau turbines were impulse types as well. They had a slightly different nozzle system. They operate well as a low pressure turbine with small drops in pressure and are used as an intermediate or low pressure turbine.

Westinghouse turbines were classified as combined impulse cross compound. They were designed as either a high pressure or low pressure turbine. They had an ahead and astern section. They had hand-controlled valves to reduce pressure in parts of the turbine for reduced load situations.

Ljungstrom turbines were a radical flow reaction turbine. It had two rotating disks to which blades were fixed. One disk rotated in a clockwise direction and the other in an anti-clockwise direction. The steam entered the center of the turbine and flowed outward.

Propellers

The propeller is such a common thing on ships and boats today that anyone not involved with this branch of engineering will have no idea how complex a subject it is. The propeller or screw as it is properly called works like a segment of a screw that pumps water from forward to aft. In theory if you create a stream of water moving in this direction it will overcome the inertia and water resistance of the ship. This water flow going past the rudder also improves steering. The subject is so complex that we can only discuss a few of the principles.

In practice there are a number of difficulties that have to be overcome for this to work in a practical manner. Some but by no means all of these problems are: vibration as the screw revolves, strength of the material the propeller is made from, size and pitch being too great for the power of the engine, slippage (the screw not biting into the water as it should) and cavitation (the screw sucking air from the surface reducing its bite in the water). The last one can be a major problem for tugs and shallow draft vessels.

In the early days the technology of the day limited the size of the propeller. A lot of these propellers had very narrow curved blades. Most were built with wide blade tips but there was a problem with vibration. This was overcome by rounding the tips. To try to overcome the vibration problem one design had a very large boss (center part) and the tips of the blades paired right down. The number of blades was another matter that was caused expermentation. Some had six blades but they were not too practical.

The other problem was that most vessels on longer voyages were also sailing ships. The propellers when not in use created a tremendous amount of water resistance. Several methods were used to try and overcome this problem until steam became reliable and economical.

The methods tried were: lift the screw out of the water, use narrow blades and feathering the blades. The methods used for removing the propeller did not work out too well. The feathering systems tried took a lot of perfecting before they became viable. One of the early attempts was to use a four bladed screw so two of the narrow blades could be lined up with the other two. The blades were then lined up with the deadwood on the stern.

The materials most used in the manufacture of propellers have been and still are: cast brass and bronze, gun metal, cast iron and cast steel. There are other more modern alloys used on some vessels today.

Some definitions:

Hub or boss - central part of the propeller.

Right hand or left handed denotes which the propeller is turning when looking forward.

Face or driving face - It is on the rear of the propeller and takes the forward thrust.

Back of the blade is on the forward side.

Leading edge - The part of the blade that cuts through the water.

Diameter - Measured to the tips of the blade.

Pitch - The same as pitch of a screw thread.

The modern screw has evolved a long way from the early attempts in this type of engineering. However somethings remain the same. The propeller must be designed for the hull and the type of work. It must also be balanced so that one blade is not heavier than the others and the pitch on all blades must match.

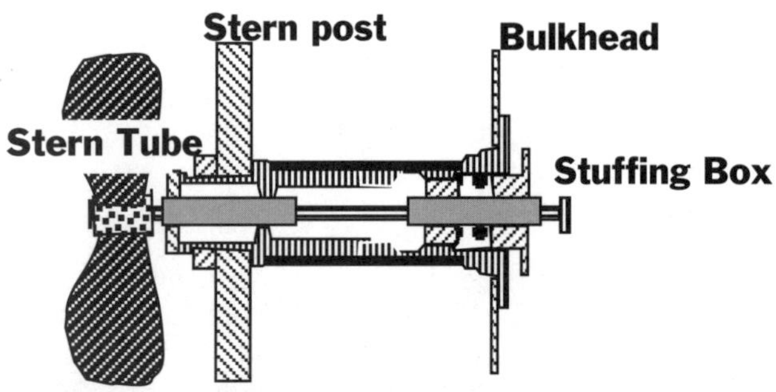

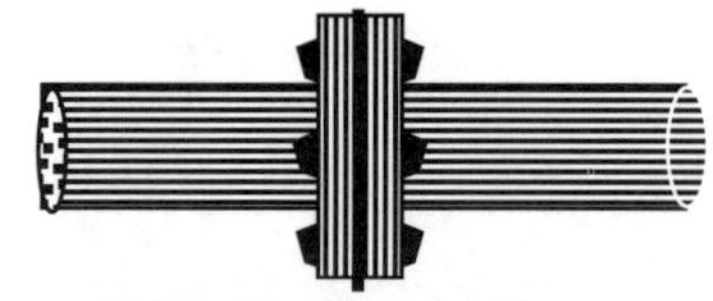

Shaft Coupling

Shafting

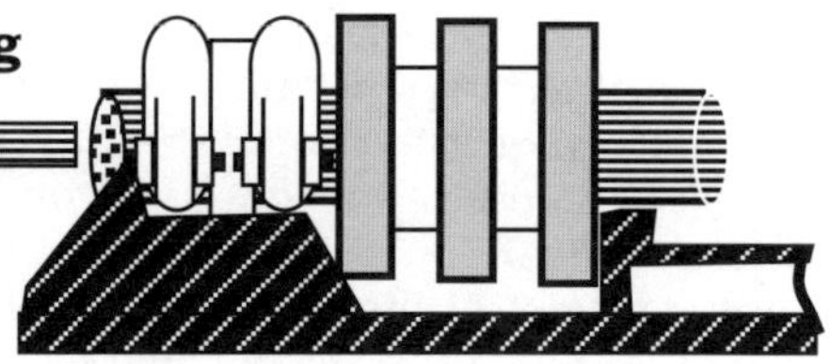

Thrust Block

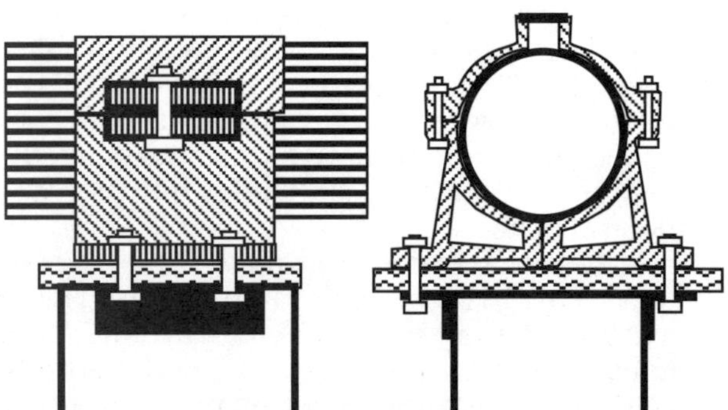

Shaft Bearing

The propeller shafts that connect the engine to the propeller must be kept in line. This was not an easy task on larger wooden vessels. The shafting is made up in sections coupled together. The tail shaft is tapered for the propeller. It usually has a brass sleeve shrunk on where it comes in contact with the stern bearing. There is a short shaft right aft of the engine that carries the work of the engine. Just aft of that the thrust bearing is located to take the thrust of the propeller. Depending on the size of the vessel and location of the engine one or more intermediate shafts are coupled to the tail shaft. Bearings located along the shafts keep things in line.

The stern tube has the lignum-vitae (hard wood) stern bearing and the stuffing box.

Auxiliary Machinery
Chapter 7

One of the advantages of steam is that there is power to any place that you can run a steam pipe or hose. At first steam ships had very little auxiliary equipment. Later steamers had a variety of equipment. Freighters had deck winches, tugs had towing winches most vessels had anchor winches. Vessels engaged in special trades may have extra winches. In addition many vessels had steam steering gear. On larger vessels deck lines were handled using steam capstans. We are only showing some of the equipment that might be found in a steam powered vessel.

Pumps

Unlike the first steamers the engine room and stokehold of any steam boat had many pumps, valves and pipes. There have been so many types of these items manufactured we can only look at a few representative types. Every steamer had feed water pumps and injectors. Bilge pumps and fresh water pumps were needed as well as fire pumps. Sometimes a pump could be used for several purposes. Just as there are many types of equipment there were many ways of handling feed water and other shipboard problems.

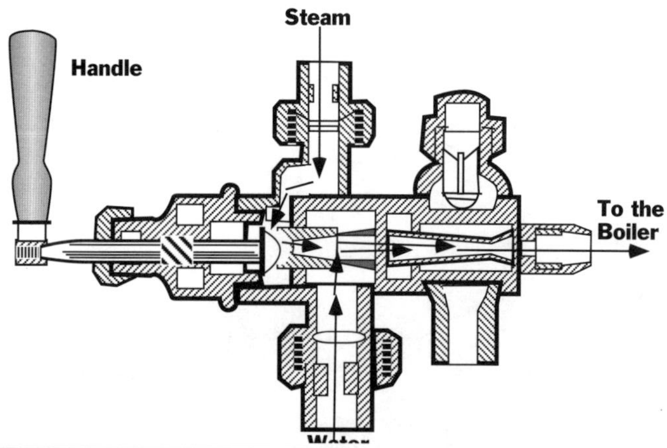

Injector -One method of putting feed water in the boiler.

Pump types:

Feed pumps

Feed pumps supplied water to the boiler. They could be operated directly from the engine or by separate pump engines. On river boats with the engines a long way from the boilers, feed pumps were usually located near the boiler. A number of vessels had both types. An injector was also used for putting water in the boiler usually from a heater.

Air pumps

Air pumps eliminate the air from the condensed steam after leaving the condenser. The vacuum created in the condenser by the condensing steam was maintained by removal of the air that came in with the exhausted steam. The air in the feed water mixed with the steam and air came from joints in the system which was released when the steam was condensed. The size and efficiency of these pumps was critical to the operation of the engine.

The air pump could be used just for air or handle both air and water. There have been many types of these pumps produced since the earliest days of steam. The single-acting vertical air pump, the Edwards Air-pump, the double-acting horizontal air pump were used for years. As engine revolutions increased so did the necessity for better air pumps.

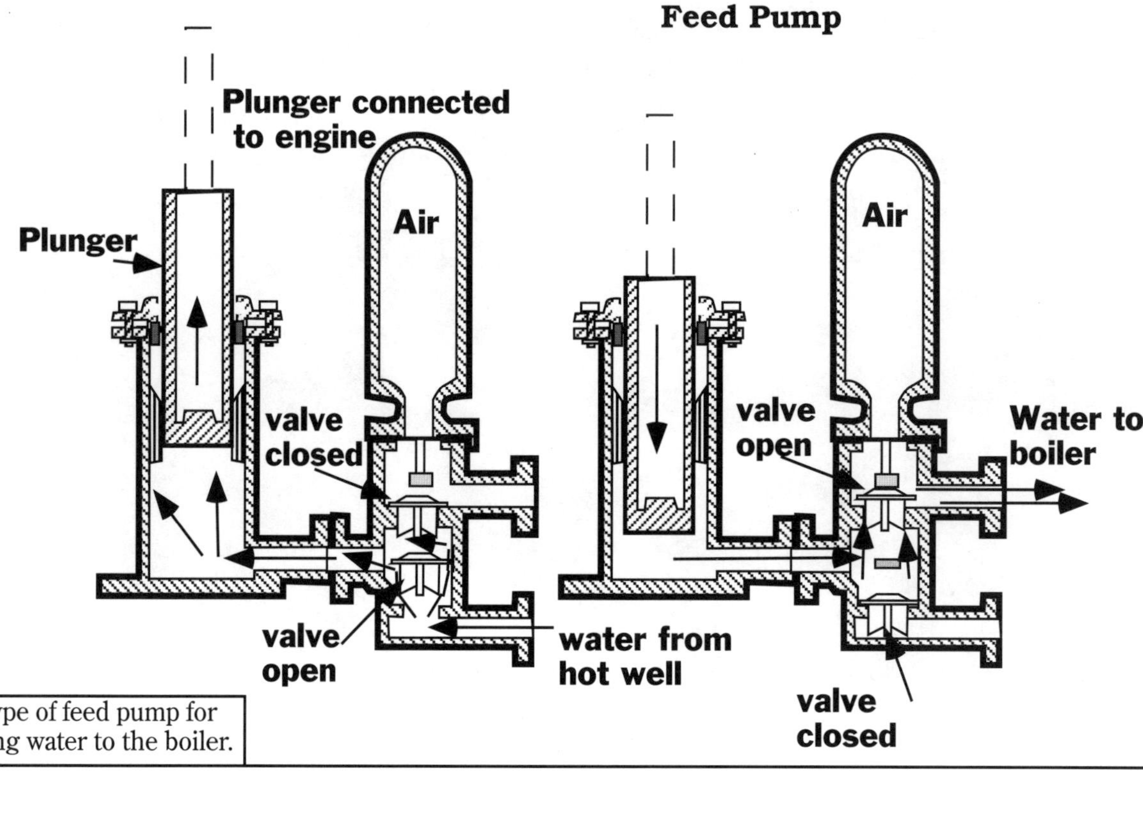

One type of feed pump for supplying water to the boiler.

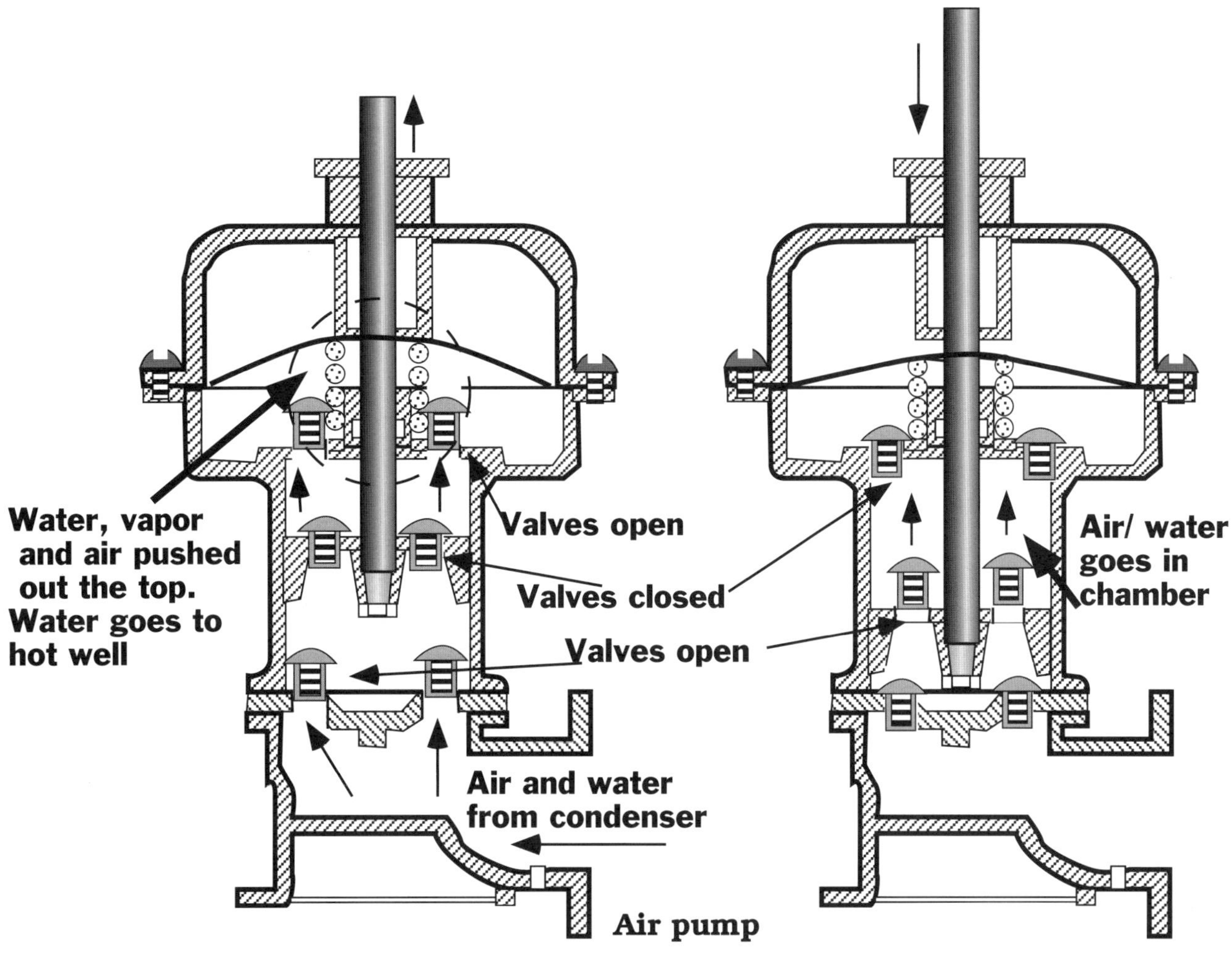

These pumps were necessary on surface and jet condensers but on a jet condenser extra water and air was handled. A separate extraction pump could be fitted to handle the water. All turbine ships use an extraction pump. The water runs or is pumped from the air pump to the hot well. Vessels without condensers such as a lot of sternwheelers did not use an air pump. Reducing back pressure on a lot of these vessels was accomplished by exhausting the steam up the stack thus reducing pressure at the exhaust valves.

On the old horizontal screw engines a horizontal-type pump was used usually under the condenser in ships. With some modification these pumps could be used on vertical engines as well.

Circulating pumps

This is a pump fitted to a surface condenser to circulate water through the tubes. The cool sea water is pumped through the tubes and back into the sea. This condenses the steam from the engine. It is usually a double-acting pump fitted with suction and delivery valves for both up and down strokes. It is often operated by the main engine but it is considered a better practice to have it run by a separate engine. This will keep the condenser cool when the engines are stopped. Circulating pumps may also supply cooling water for lubricating oil and drain coolers. They can double as bilge pumps in an emergency.

Air vessel

An air vessel is a cast iron dome fitted to some pumps to receive the air forced into it by the water on the up and down stroke. The air acts as a cushion creating a steady flow and preventing heavy knocks in the pipes.

Duplex pumps

A duplex pump is a double-acting pump and is one in which the steam and water cylinders are in duplicate. They can be vertical or horizontal. An attribute of these pumps is that they deliver an almost uniform flow of water. When one plunger has reached the end of its stroke the other is still moving. No dead center exists as one or the other of the steam valves is always open. They are used for many purposes including feed, fire, ballast, sanitary and general service. They can be vertical or horizontal pumps.

Centrifugal pumps

A centrifugal pump has a wheel and impeller with vanes as arms which act on the water to give it a circular motion. It is enclosed in a chamber into which the water flows and is forced out the other end by the impeller. Used for many different purposes such as a circulating pump for a surface condenser.

Flywheel pumps

On early ships flywheel pumps were common. This pump consisted of horizontal steam and water cylinders with the piston and plunger on a common rod and moved together. Attached to the rod was a crosshead with a connecting rod leading to a crank and shaft carrying a flywheel. The flywheel served to carry the pump past the dead points. The shaft had an eccentric to operate a slide valve.

Direct-acting pump

This common pump is horizontal and has been produced by a number of companies with slight differences in the pump and steam engine. It has two cylinders one for steam and one for water. One piston rod operates both ends. The steam end is a simple reciprocating engine. The pump has inflow and outflow valves on each end of the cylinder. It is a double-acting pump.

As an eccentric cannot be used a slide valve is operated by an auxiliary piston. An auxiliary valve controls the exhaust and steam for the piston. A linkage from the piston rod works the auxiliary valve. Other versions of this pump use a slightly different method of operating the slide valve. When two of these pumps are hooked up side by side they are a duplex pump. On a duplex pump the piston rod operates the valve gear of the other pump.

Gear type oil pump

This pump consists of two small toothed wheels gearing with one another and fitting exactly into a casing. When running, the oil is moved around between the wheels and the case. There are no suction or delivery valves to create problems. Used for pumping oil.

Doctor

On western rivers' vessels and probably on some of the West Coast sternwheelers a combination feed pump and feed water heater was used. The doctor was a small vertical beam engine with a crank and flywheel operating four pumps. Two of them were lift pumps that took water from the river and delivered it to heaters overhead. The other two were feed pumps that delivered water from the heaters to the boilers. As these vessels were usually non-condensing, these pumps had to supply all feed water to the boilers.

Condensers

In order for a steam engine to operate the pressure must be lower on the exhaust side than on the inlet side. This is usually done by condensing the exhausted steam and creating a vacuum. This was extremely important on the old low pressure engines. On marine applications reusing the fresh water is of prime importance. This could be done with a surface condenser but not a jet condenser.

Jet Condenser

The first type of practical condenser was the jet condenser. Cold water was mixed with the steam to cool and condense it. The water was sent back to the hot well while the air was pumped out. The big disadvantage of these condensers was that salt water was introduced into the boiler. They were used for years. Engineering manuals printed around the turn of the century still allude to them.

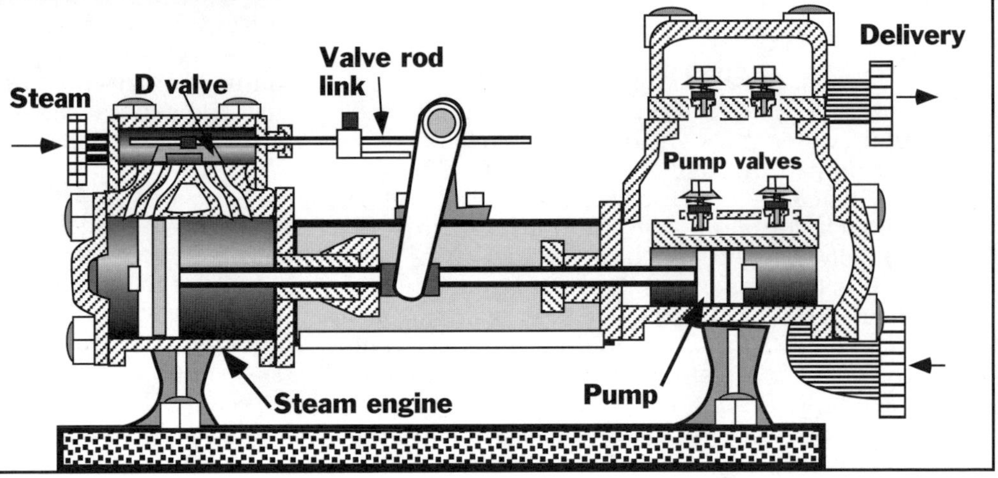

These direct-acting pumps are found in most engine rooms.The pump is operated by a small reciprocating engine. When two of these pumps are hooked together they are called duplex pumps. This is however not the only type of duplex pump.

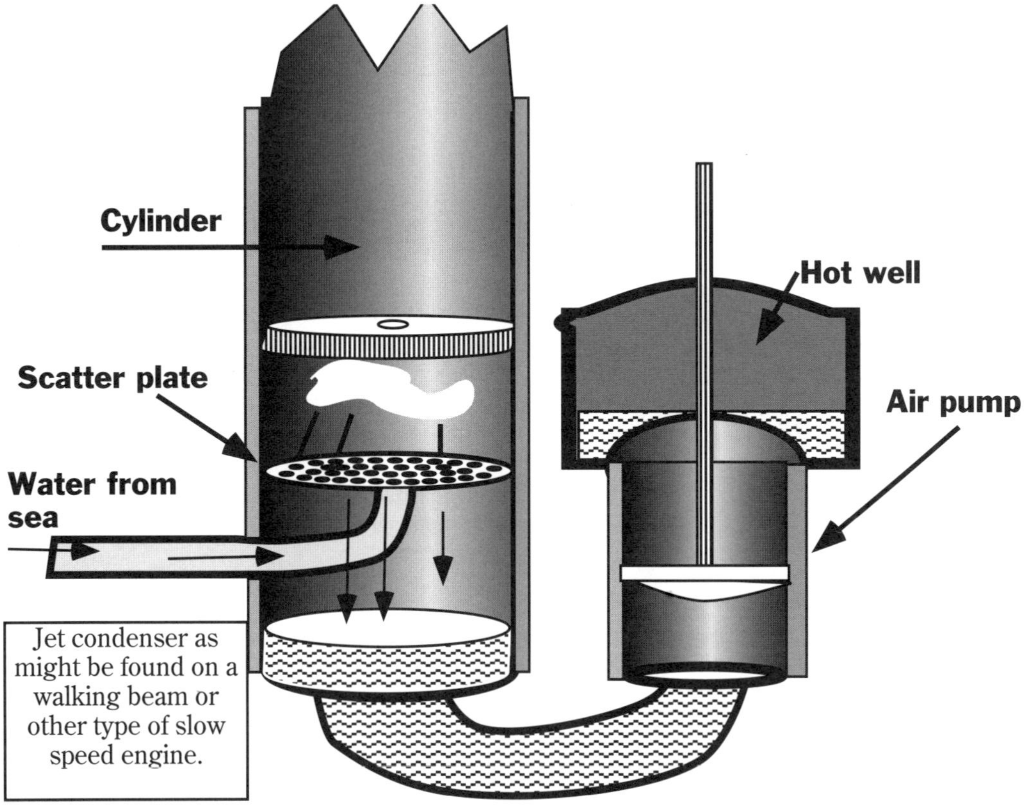

Jet condenser as might be found on a walking beam or other type of slow speed engine.

Jet condenser from small tug in Bellingham.

The jet condenser consisted of an air tight chamber into which the exhausted steam flows. The passage of the steam was intercepted by a spray of water created by the inrush of water through small holes or narrow slits in a pipe placed across the streamway. If the spray was small like rain it would mix mechanically with the steam and cool it with surface contact. If slits were used in the pipe, cooling the steam was by surface contact only. The steam was turned to water which fell to the bottom and was pumped to the hot well by the air pump. In order to create a vacuum the air pump pulled out the air. On some condensers common in North America, the water was led into the top of the chamber and let fall on a scattering plate. The scattering plate had a number of small holes which created a spray of water over the steam. The steam was let into the bottom of the condenser. This type was used on a lot of early walking beam engines.

The shape of the jet condenser varied a lot over the years. It was immaterial to its operation as long as the inlet steam was high enough to prevent water running back into the cylinder. The bottom was formed so that the water would drain into the bottom for removal by the air pump. It was generally formed to suit the shape of the ship and machinery on early ships.

The capacity of the jet condenser should not be less than one-forth that of the cylinder or cylinders exhausting into it. Usually one-third was the figure used. Too small a condenser would flood an overflow into the cylinders.

The amount of cold water injected had to be regulated so the water going into the hotwell was not too cool. As the feed water came from the hotwell it needed to be as warm as possible. The engineer adjusted the vacuum to around 24 inches, and the temperature in the hotwell to between 110°f (43.3c.) and 120°f (48.9c.), though a careful engineer could get 130°f (54.4c.).

A second injection valve was fitted to the bottom of the condenser for picking up water from the bilge. This was often used in wooden paddle wheelers to save the ship.

A non-return snifting valve was often installed in the bottom of the condenser for blowing out water with steam. Excess water that builds up was pumped over the side. All this resulted in a salt water to fresh water ratio of about 30 to 1 which created corrosion problems.

Surface Condensers

Sea water was forced through tubes in the condenser by the circulating pump. The water was forced back and forth through the tubes on some models. Steam exhausted from the LP cylinder was brought to the top of the case through the eduction pipe. The vacuum gage was located on the top as well. The air pump suction was on the bottom. This pulled the steam passed the tubes where it condensed. Water ran to the bottom where it was removed by the air pump to the hotwell.

The first really practical surface condenser was patented by Samuel Hall in 1831. It was not until 1860 that the surface condenser started coming into general use. It was not until mineral oil replaced tallow for lubrication that it would work successfully. Part of the trouble with Hall's con-

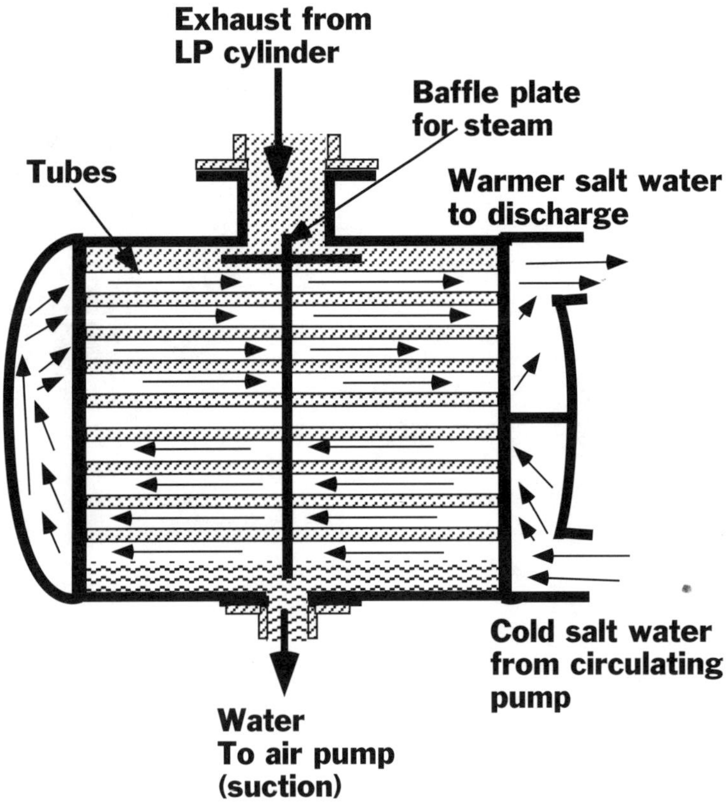

Surface condenser
The cold salt water enters the bottom of the condenser from the circulating pump. After going back and forth through the tubes it is discharged over the side.
Steam from the LP cylinder is drawn into the top of the condenser by the vacuum created in the condenser. It condenses on the cold tubes as it is pulled down by the air pump. The mixture of water and air goes out through the bottom to the air pump. The warm water goes to the hot well after the air is extracted.
There are more tubes in a condenser than shown in the diagram. The condensers are built in a number of ways but the concept is the same.

denser was that steam went through the tubes with the cooling water circulating outside them. If the cooling water had gone through the tubes with the steam on the outside it would not have plugged with the tallow and other junk in the steam. Other problems with metals used in the construction had to be overcome as well.

The surface condenser consisted of a cast-iron chamber, or cylindrical vessel made out of boiler plating. Some were constructed out of cast brass. They were generally placed at the back of a reciprocating engine. Brass tube plates were placed at each end. The tubes running between these plates were made of brass, Muntz metal, Monel metal or Cuprous nickel. Larger ones had a supporting plate in the center. Stay bolts were fitted on some of them. The tubes had packing to keep them tight. Doors were fitted at the ends for inspection and repair.

There still can be problems with lubricating oils and dirt from operating in rivers or shallow water. The condenser may have to be flushed out with fresh water and soda.

The hotwell temperature is worked at around 120f° (48.9c.) to 140f°. (60c.). With a lower temperature the vacuum would be increased, but if the feed water is too cool fuel consumption goes up and it can create problems in the boiler. The amount of vacuum will vary with fluctuations in the barometer and the temperature of the sea water. There are filters placed to collect as much oil as possible before the water is returned to the boiler.

On newer ships a closed system is in use. The condensate is withdrawn by a rotary extraction pump and fed to the suction side of the feed pump. Air is removed by an air ejector. This only works when the main engines are running.

Feed Heaters

As we have described before one of the engineer's problems has been to try and keep a good vacuum while keeping the feed water temperature up. One of the ways of overcoming this is to use a feed water heater between the hotwell and the boiler. Either steam or waste furnace gases can be used in the heater. If waste gases are used in a watertube boiler it is part of the boiler. There are several types of steam heaters using steam from the main or auxiliary steam pipe or from the exhaust of some of the auxiliaries on the way to the condenser.

Evaporators

Evaporators are used to manufacture fresh water at sea to make up for loss of feed water. They are made up of a series of nests of coils of pipe contained in a chamber. The chamber has a salt water inlet and the steam is passed inside the tubes. The

Air pump and condenser Virginia V. On some small vessels the air pump and circulating pump were operated by separate engines. This allowed exhaust steam from winches and generators to go back to the condenser when the main engine was not running.

heat in the steam passes through the tubes and forms steam or vapor on the salt water side. The chamber is connected to the condenser or low pressure receiver and the steam formed in the evaporator goes into the feed water after it is condensed. Evaporators are not too common on coastal vessels. As with most steam equipment there were several types of evaporators in use.

The circulating pump and main condenser of a Liberty ship. This is a surface condenser.

Capstan on the sternwheeler Sicamous
The stern wheelers used different types of capstans or winches. If the vessel was a tug it would probably have one or more multiple drum winches on the fore deck. The engine for a capstan is located below the deck. It could be just under the deck or further down in the ship where it was easier to maintain.

Winches

Every steamer used capstans and winches of some type. The anchor winch or windlass is used for handling mooring lines as well as the anchor chain. The wildcats for handling the chain have a clutch system so they can run free when running out cable. The gypsies (warping ends) can be used for handling lines. They are also fitted with brakes for each wildcat. The direction is usually handled with one lever and speed with a valve. Most windlasses have two cylinders.

On larger tugs capstans are often found on the after deck for handling gear. At one time they were used for handling towlines but this is now rare on this coast. Capstans with a wildcat can be used for anchor work. The engine is under the deck. Controls are led up through the deck.

Towing winches can be found on most West Coast tugs. Some of them are very large with several drums for the towline and strawline others are simple one drum winches. At one time the smaller ones did not have spooling gear for the towline. This was done by a seaman forcing the incoming line over with an iron bar. The spooling gear operates on an endless worm gear connected to a gear train from the drum shaft.

Some steam towing winches could be set so that when a strain came on the towline, such as the tow hiting a big sea, it would payout line and when the strain came off the towline the winch would heave it in again. If the tug did not have this feature

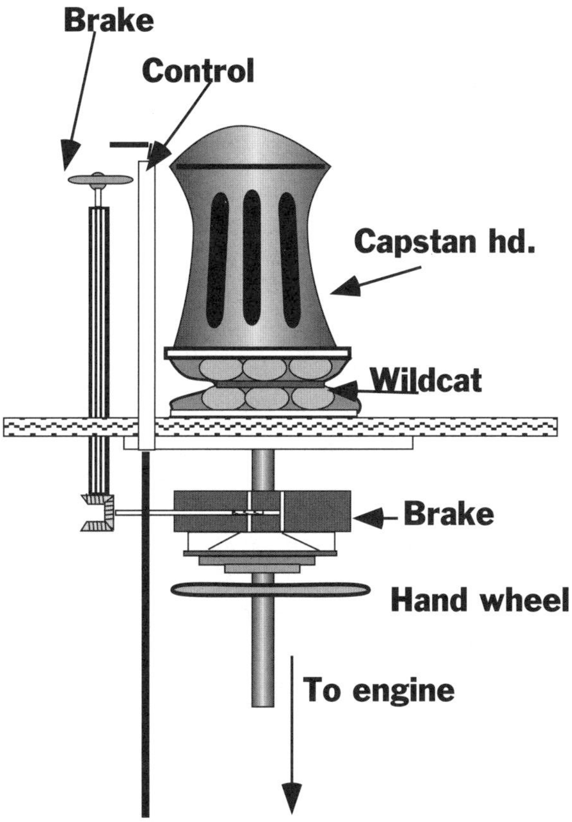

Towing winch on the S.S. Master. This type of steam winch was used on hundreds of tugs on the Coast. This one has spooling gear, operated with a hand wheel for spooling the towline on the drum. On many winches the spooling gear was run by a worm gear connected to the drive. On some of them spooling was done by using a crowbar jammed into the base to force the towline over. Large ocean -going tugs could have two drums. There was a clutch for the drum, a drum brake, a lever for heaving in or out and a throttle.

it would be done by hand. When working on a relatively short towline like crossing the Columbia River Bar, the towline is eased in and out by one of the mates.

Cargo winches are operated with a single lever for up and down and speed. A foot brake is used. On coastal vessels one man operates two winches while on deep water ships there is a man for each winch. Lifting the lever takes in the runner and lifts the cargo. Lowering the handle pays out the runner and lowers the cargo. The author worked on one ship where the winchman operated in reverse, down for up and up for down.

Most winches have one thing in common: the engine operates a small gear at a relatively high speed which turns a large bull gear. Some of the engines have two eccentrics while others have only one.

Reversing with one eccentric

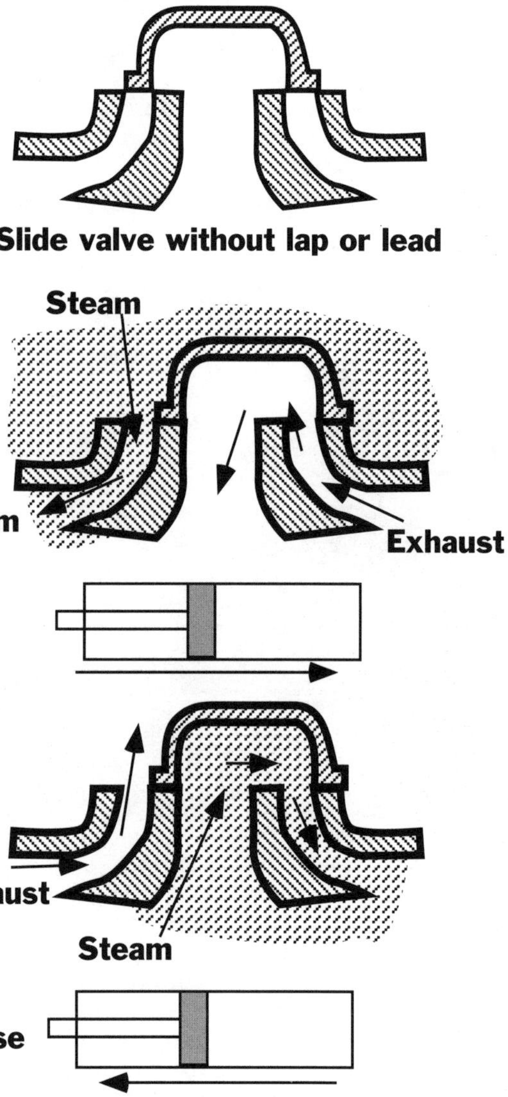

Many small engines such as those used on winches had one eccentric. Reversing was accomplished by using a valve that had no lap or lead. The valve was switched from an outside valve to an inside valve. The valve had to be long enough to cover both sides of the steam ports at the same time. There was no cut-off or cushioning in these small engines. The valve opens to steam at one end of the cylinder at the same time it opens to exhaust at the other. To reverse the engine a valve changes the steam outside of the valve (outside valve) to the inside (inside valve). This lets the steam enter the port at the other end of the cylinder and reverses the engine. The same valve that switches the steam can also work as the throttle. This makes the engine controllable with just one lever.

The winch on the sternwheeler Sanson V. As a snag boat she needed a lot of power for her derrick. Each drum had a friction clutch and a foot operated brake. The throttle was operated with a lever.

The windless on a deep water ship. Smaller versions were used on ships of all sizes. Each wildcat had a separate clutch and a brake. Two cylinders on opposite sides of the winch turned a shaft with one or small gears on it. The small gear engaged a large gear or gears connected to the main shaft. This shaft had a gypsy on each end.
A lever controlled the direction the shaft turned. The speed is controlled by a separate valve.

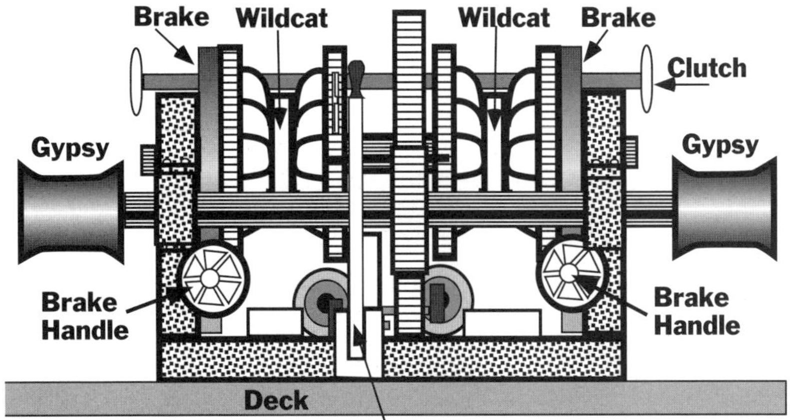

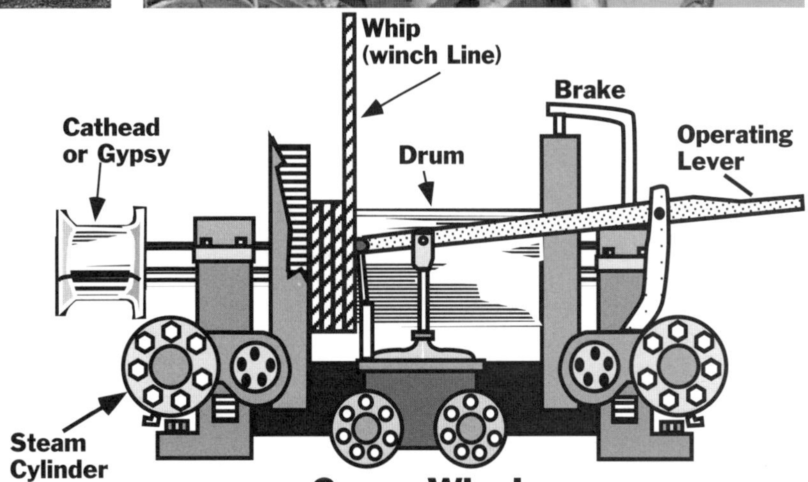

Cargo Winch
Reverse-valve Steam Winch

Cargo winches. Had a clutch from the drum to the shaft. When operating the direction and speed of the winch was operated with a single lever. Most of the winch engines had one eccentric.
On coastal ships one man worked two side-by-side winches with a lever in each hand.

Coastal Winch Setup

Dynamo Electric light was introduced to ships in the 1890's. The arc light was used in the 1880's.

Electrical

It is hard to comprehend operating a vessel without any electricity on board but for years there was no electric light on steamers. Dynamos were eventually installed in all vessels. This brought in searchlights and electric navigation lights. Electric lights replaced lamps in passenger and crew spaces.

The dynamo is run by a separate engine. The output was usually DC. If the engine was running the exhaust would go to the condenser. On smaller vessels such as tugs this meant the vessel reverted to lamps if the ship was tied up. Passenger vessels had a smaller auxiliary condenser for running equipment in port.

Steering gear

The first steamships had the same type of manual steering gear as the sailing ships. With screw-driven ships the effect of the propeller wash on the rudder made steering more difficult. The steering was moved from the stern to the wheelhouse often near the center of the ship. This meant a series of rods and chain connecting the wheel to the rudder had to be installed.

By using enclosed rods running along or under the deck, wire to the quadrant and wire or sometimes rope to the wheel, reliable steering could be retained. Emergency steering gear was still located aft. This type of gear lasted on some tugs and other smaller vessels to the end of the steam era.

In heavy weather or when manoeuvering, this gear made the wheel hard to turn. For this reason steam steering was developed. There have been a lot of designs. All of them use steam to assist or actually move the rudder stock. Steam steering engines were often slow moving and could be noisy which was a problem in a strong current or bad weather.

Some designs had the steam steering engines directly under the wheelhouse. The machinery operated wires or shafting leading back to the quadrant. Others had the machinery in the engine room with connections to the quadrant. There were some geared directly to the quadrant and operated with rods from the wheel. A hydraulic telemotor was used on some later ships to operate the steering engine. Telemotor control was patented in 1888 by P.B. Brown. This allowed the operation of the control valves of the steering engine from a distance without the use of rods or chains. Hydraulic and electric systems have been used for years.

Steam steering engines usually have two pistons. Most of the them fitted with three valves, a central control valve, and two piston valves. The control valve distributes the steam to the engine valves so the gear can run to port or starboard.

All passenger and most other commercial vessels are required to have an alternative method of steering if the main one gives out. Some of the emergency steering gear for tugs works better on the drawing board than on the vessel.

All the first steamship rudders were of the hinged or blade unbalanced type. Steam tugs used an unbalanced or a form of semi-balanced blade rudder. A tug has a large rudder for the size of the vessel as it needs a lot of lift (turning power) for handling tows. Larger vessels could be fitted with semi-balanced or balanced rudders. These rudders could be an areofoil type.

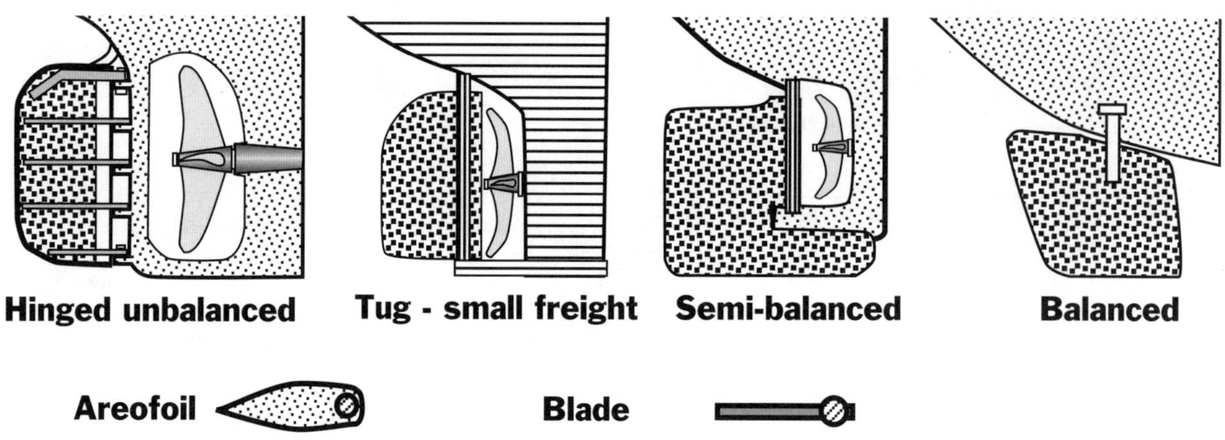

Bibliography

A Catechism of the Steam Engine 1857 D. Appleton & Co.,Reprint, Manual, J. Bourne, New York Reprinted 1878 illustrations changed

A Manual of Marine Engineering 1904 Charles Griffin and Company LTD., Manual, A.E. Seaton London, 15 edition (org 1883)

A Pacific Legacy 1991 Chronicle Books, Photographic, Wayne Bonnett, San Francisco

Against Wind and Weather 1981 Douglas & McIntyre, History, Ken Drushka, Vancouver

Alaska Steam 1984 Alaska Geographic Society, Photo History, Lucile McDonald, Anchorage

American Steamships On The Atlantic 1981 Associated University Press, Technical, Cedric Ridgely-Nevitt, East Brunswick - New Jersey

Audels New Marine Engineers Guide 1918-37 Theo. Audel & Co., Manual, Theo Lucas, F.D. Graham New York

B.R. 77 1941 Kings Printer, Manual, HMC Ottawa

Bailey's Handbook 1932-53 Goodhart-Wllcox Co., Manual, A.R.Bailey, Chicago

Breakdowns - Reed's Hints to Sea-going Engineers 1903, Manual, Thomas Reed & Co., Sunderland, London

Encyclopedia of Nautical Knowledge 1992 Cornell Maritime Press, Marine Encyclopaedia, W.A. McEwen & A.W. Lewis, Centerville Maryland

Exploring Shipwrecks 1988 Archeology, K. Morris - P. Rowlands, London

Historic Ships of San Francisco 1984 General marine, Chronicle Books, Steven E Livingston, San Francisco

ICS Steam Engineering 1945 International Textbook Co., Course, ICS -USA - England

Merchant Ship Types 1975 Marine Media Management Ltd.,Technical, R.Munro-Smith, London

Modern Marine Engineer's Manual 1941 Cornell Maritime Press, Manual, Alan Osbourne, New York

New Catechism of The Steam Engine 1904 Theo. Audel & Co. Publishers, Manual, N. Hawkins M.E.,New York, Org 1897

Nicholl's Seamanship & Nautical Knowledge 1945 Brown, Son & Ferguson Ltd., Manual Charles H. Brown F.R.S.G.S., Glasgow

Pacific Coastal Liners 1959 Bonanza Books, Photographic, G. Newell & J. Willamson, New York

Paddle Wheel Days In California 1946 Stanford University Press, General history, Jerry Mac Mullen

Paddle Wheel Steamers and their Giant Engines 1987 Seamaster Inc., Technical history, Bob Whittier Dusbury Massachuetts

Paddlewheelers on the Frontier 1967 Foremost Publishing Ltd.,history, Art Downs, Surrey B.C.

Practical Marine Engineering 1901 Marine Engineering Inc., Manual, William F. Durand, New York

Prince Ships Of Northern B.C. 1995 Heritage House Publishing Co., History, Norman Hacking Surrey B.C.

Research in Steam Engineering 1865 William Hamilton, Hall, Franklin Institute, Technical B.F.Isherwood USN, Philadelphia

S.S. Great Britain The Model Shop 1982 Argus Books ltd., General, William Mowll, Hetfordshire GB

Ships and Shipwrecks of the Americas 1988 Thames & Hudson, Archeology, George E Bass, London

Ships Monthly 1969 - 70 Twentieth Century Press, Magazine, London

Ships of the Inland Sea 1960 Binfords & Mort, History, Gorden R. Newell, Portland Oregon

Steam at Sea

Steam Navigation 1898 1898 reprint, William Briggs, Montreal News Co., General, James Croil, Montreal

Steamboat Virginia V 1985 Documentary Book Publishers Corp., History, M.S. Kline

Steamboats Come True 1978 Little, Brown, & Co., General, James T. Flexner, Boston Toronto

Steamboats on the Western Rivers 1993 (1949) Dover Publications Inc., Academic, Louis C Hunter New York

Steamships and Motorships of the W.Coast 1968 Superior Publishing, Photo History, Richard M. Benson, Seattle

Steamships and Their Story

Stern Wheelers Up The Columbia 1977 University of Nebraska Press, History, Randall V. Mills

Strenwheelers and Steam Tugs 1984 Sono Nis Press, History, Robert D. Turner, Victoria B.C.

Textbook on Marine Engineering 1900 I.C.S., Text Book, Colliery Engineering co., Scranton U.S.

The Beaver 1993 Horsdal & Schubart Publishers Ltd., General Marine., James P Delgado, Victoria B.C.

The Marine Steam Engine 1918 Longmans Green and Co., Manual, R. Sennett & H.J. Oram, London

The Mississippi Steamboat Era 1987 Dover Books, Photo, Joan W. Gandy & Thomas H. Gandy,

Mineola N.Y.
The Oxford Companion to Ships & The Sea 1979 Oxford University Press, Marine Encyclopedia, Peter Kemp, London - Glasgow - Oxford
The Panama Route, 1848-1869 1990 University of South Carolina Press, History, John H. Kemble, South Carolina
The Princess Story 1974 Mitchell Press Ltd., History, Norman Hacking & W. Kaye Lamb, Vancouver B.C.
The Running & Maintenance of Marine Machinery 1965 The Institute of Marine Engineers, Manual, London, 5 edition
The Sea Remembers 1987 Smithmark Books, Archeology , Peter Throckmorton, N.Y. N.Y.
The Ship 1980 Her Majesty's Stationery Office, History, John M. Maber, London
Time, Tide & Timber 1949 Stanford University Press, Histort, E.T. Coman, Jr. & H. M. Gibbs
Tod & McGibbon's Questions & Answers for Marine Engineers 1940's? James Munro & Company, Manual, Tod & McGibbon, Glasgow, 10 edition
Trillium and Toronto Island 1941 Firefly Books LTD. General, Mike Filey, Willowdale Ontario
Use And Abuse of The Steam Boiler 1876 Keystone Press, Manual, Stephen Roper, Philadelphia 6 edition
Verbal Notes & Sketches-Manual of Steam Engineering. 1940's to 1950's James Munro & Co., Manual, J.W.M. Southern, Glasgow, 12 -17 - 18 editions
Where Go the Boats Peace River Centennial Museum, booklet, Evelyn Hansen, Peace River Alberta
Whistle Up The Inlet 1974 JJ Douglas Ltd., History, Gerald A. Rushton, Vancouver B.C.
Yukon River Steamboats 1982 Pictorial Histories Publishing Co., Photo, Stan Cohen, Missoula Montana.

Definitions

Air heater A device for preheating the air for combustion
Air pump A pump that extracts air and condensed water from the condenser and discharges it into the hot well.
Beam engine A beam is used to transfer the action of the piston to the crankshaft.
Bed plate Heavy iron casting forming the base of a marine engine. This is also called a sole plate.
Boiler - cochran's Early type of watertube boiler.
Boiler - double ended A double ended scotch boiler has a furnace at both ends.
Boiler - drop-flue A boiler whose smoke and combustion products go down before returning through the flues to the uptake and stack. This is a return-flue boiler.
Boiler - dry bottom or wet bottom Rectangular boilers were either wet bottomed or dry. The wet bottomed had independent furnaces inside the shell surrounded by water. Dry bottom furnaces were united at the bottom.
Boiler -dryback There is no water leg between the combustion chamber and back head.
Boiler- express water tube boiler Small closely spaced tubes.
Boiler - fire tube - A boiler that the smoke and combustion gas goes through tubes to heat the surrounding water.
Boiler Flues Large diameter passages between a boiler furnace and the uptake.
Boiler- large tube water tube boiler The tubes are large and straight.
Boiler - Rectangular An early type of boiler used in European ships. Stays ran athwart ships as well as fore and aft.
Boiler- Return tube boiler Similar to a return-flue boiler except that many small tubes are used rather than the bigger flues.
Boiler - Return-Flue Boiler A boiler whose smoke and gases flow away from the grates then reverse direction and return through the flues before going up the stack.
Boiler scale Mixture of Carbonate of Lime, Sulphate of Lime, Magnesia and traces of salt found on heating surfaces.
Boiler shell The thin, hollow cylinder forming the outer envelope of a boiler (circular boiler) consisting of one or more plates joined together.
Boiler- Triple pass boiler The smoke and gases go back and forth three times before going up the stack.
Boiler - wet - back There is a water leg between the combustion chamber and the back head on a scotch boiler.
Circulating pump Circulates sea water through tubes in surface condenser.
Compound engine An engine where the steam is expanded in two stages. First in the HP cylinder then in the LP cylinder.
Condenser jet (jet condenser) Steam is condensed by spraying in cold sea water.
Condenser surface (surface condenser) Steam is condensed by coming in contact with cold tubes filled with sea water.

Condensing engine When low pressure engines were used they needed a condenser to operate. The so called H P engines did not.
Crosshead Connecting part between piston rod and main (connecting) rod.
Crosshead guide Piece of hard smooth-faced metal fitted on the inside of a cylinder column or engine frame on which the shoe of a crosshead slides.
Crosshead links Connecting links between a crosshead and a pump lever.
Crosshead nut Nut which secures a piston rod to a crosshead.
Crosshead pin Pin connecting the connecting to an engine crosshead.
Crosstail Part of a side-lever engine connecting the two side levers at their ends farthest from the cylinder. From its center a connecting rod turns the crank on the paddle shaft.
Cushioning The amount of steam left in a cylinder and compressed by the piston at the end of the stroke.
Cut-off valve Part of the valve gear that regulates the period during which steam is admitted to a cylinder.
Cutting off steam The slide valve has closed the steam port allowing the steam in the cylinder to expand for the rest of the stroke.
Cylinder escape valves Safeguard against accidents through over pressure or water accumulating in the cylinders. Placed on the top and bottom of IP and LP cylinders.
Direct-acting engine Steam engine where the piston acted directly on the crank shaft often with a connecting rod. No beam or other intermediate device used.
Doctor Small vertical beam engine used to pump water to heaters and pump feed water. Used on western river boats.
Domes Domes are superstructures to boiler shells attached by riveting or other means.
Domes (steam domes) Steam domes on marine boilers are often used to increase the steam space. Found on small fire tube boilers of the fire box, horizontal return-tubular, and flue type.
Dry pipes Installed inside boilers to reduce priming.
Duplex pump A pump consisting of two steam cylinders and two pumps. Gives steady stream of water. Can be vertical or horizontal.
Economisers An economiser consists of a series of tubes through which feed water passes on the way to the boiler to warm it.
Feed pump A pump drawing condensed water from the hotwell and forcing it into the boiler. Engines without a condenser may take water from another source.
Feed water Water used to keep the water level up in a boiler.
Flues Cylindrical shells for providing additional heating surface and to form a path for product combustion.
Full Stroke cam On Western Rivers' stern wheeler- Cam that lets steam in for the full stroke on the horizontal engine. (poppet valves)
Furnace The fire box or large flue in which fuel is burned.
Gas baffles Gas baffles are installed in some boilers to change or alternate the path of the combustion gases.
Header Headers serve as front and rear water spaces and are manufactured in shapes into which tubes can be rolled.
Horizontal engine Cylinder nearly horizontal. Two types the low pressure used on early screw-driven vessels and the high pressure engine used on the stern and sidewheelers.
Hot Well Tank to which the air pump discharges water from the condenser.
Indicated Horsepower The measure of work done in the cylinder of a steam engine. IHP= area of piston in inches x psi x number feet travelled by the piston per minute / 33,000.
Inverted directing engine On early engines the pistons were under the crankshaft. When the cylinders were placed above the shaft for screw-driven vessels, the engines were called inverted. This was dropped when most reciprocating engines had the cylinders above the shaft.
Lagging Covering of boiler and steam pipes to prevent loss of heat. Usually asbestos.
Lap or cover The steam lap is the distance the valve covers the steam edge of the ports when in mid position.
Lead - The amount the valve is open at the end of the piston stroke to give a cushion of steam.
Lever Engine An engine where the valves were operated by levers and wipers such as the western rivers HP horizontal engine.
Main engine column Main engine column is a pillar for supporting the cylinders in a vertical reciprocating engine.
Mud drums, Mud-drums are attached to boilers to provide a quiet place for the collection of mud and sediment in mechanical suspension in the feed water.
Multiple expansion engine Early term for compound or triple expansion engines.
Nominal horsepower Used in builders' ratings of engines. A leftover from the early days. There are several ways of calculating it. Board of Trade method used by governments.
Paddle crank Arm on paddle shaft of a stern wheeler that turns the paddle wheel.
Pitman arm The long arm connecting the crosshead to the paddlewheel crank on a horizontal engine.

Poppet valve Valve consisting of a stem, and a circular top that is lifted and lowered in a seat. Sometimes used on steam engines.
Priming Water mixed with the steam going from the boiler to the engine.
Quadrants control (control quadrants) Quadrants where the control levers are located on a sternwheeler.
Range of temperature Difference between the temperature of commencement of the stroke and the exhaust temperature at the end of the stroke.
Reciprocating engine An engine in which to and fro or backward and forward motion is employed. The pressure of the steam moves the piston from one end of the cylinder to the other. A crankshaft is used with a connecting rod to the piston rod.
Smoke box Airtight space at the end of boiler where waste combustion gases are sent to the stack. Also known as uptake.
Sole plate Plate engine secured to. Sometimes called the base plate.
Stay bolt hollow (hollow stay bolt) Hollow threaded bar screwed through plates.
Stay bolt solid (solid stay bolt) Threaded bar screwed through the plates-boiler.
Stay through (through stay) Threaded solid bar extended through both heads of a boiler.
Steam drum Serves as a steam collector and retainer.
Steam -saturated (saturated) steam Steam directly from boiler. Steam in direct contact with the water from which it is formed.
Steam superheated (superheated steam) Steam that has been reheated to get the water out of it. It is hotter.
Steam wet (wet steam) Steam and water mixed due to priming taking place in the boiler.
Stephenson Link Common method of determining which way the engine will run, ahead or astern, using eccentrics.
Superheaters Any arrangement of heating surface designed to add heat to steam, after the steam has been separated from the boiling water.
Three quarter cut-off cam Western rivers boats. The cam can cut off the steam after three quarters of the stroke. The cam is changed after the vessel is going full away.
Thrust bearing Bearing to take the thrust of the propeller.
Torque Turning movement to which a shaft is subjected when a force is extended to rotate the shaft against a resistance.(turning force on a shaft).
Triple-expansion The steam is expanded three times.
Tubes - boiler (boiler tubes) Tubes are cylindrical shells of comparatively small diameter constituting the main part of the heating surface of a boiler.
Valve - lead Lead is the amount by which the valve is open when the piston is at the end of the stroke.
Valve - piston Piston shaped valve used for regulating steam and exhaust in the cylinder.
Valves - D type D shaped valve that regulates exhaust and steam on one cylinder.
Vertical direct-acting engine The pistons, supported on columns are above the crankshaft. On early engines several cylinders acted as separate engines unlike the compound engine.

Index

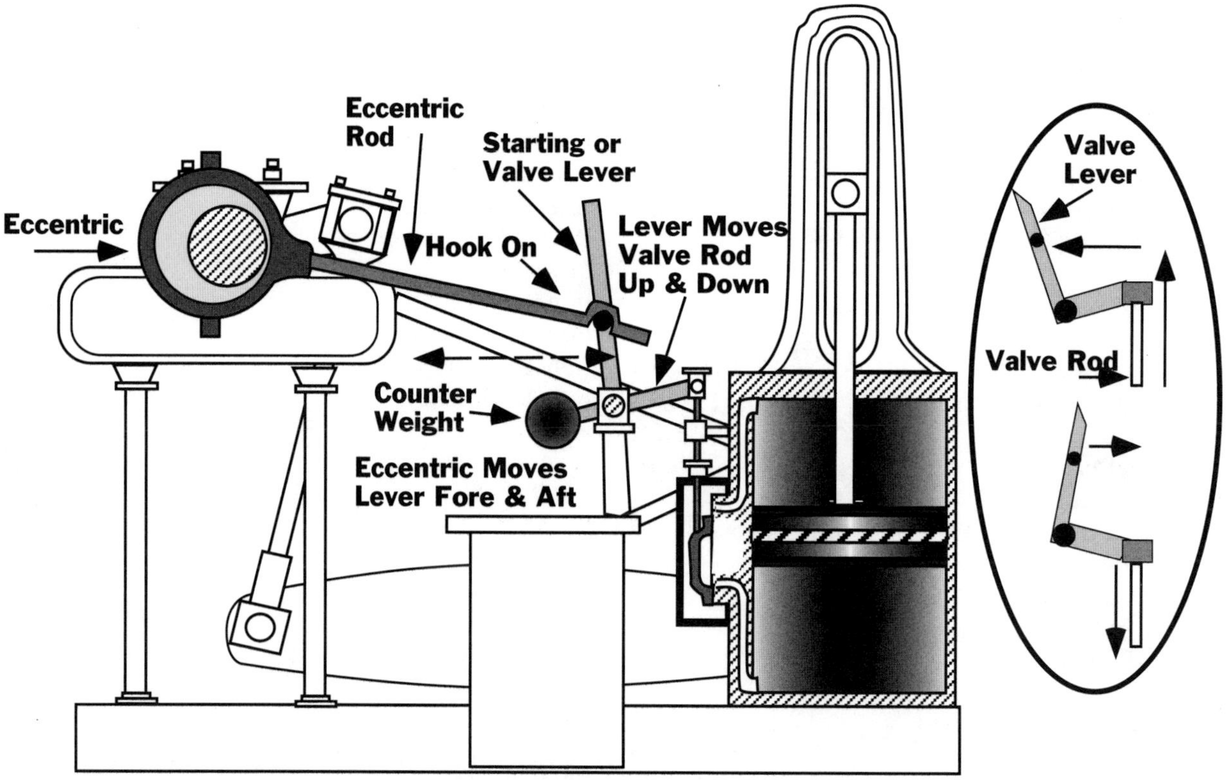

The valve system on a simple side-lever engine. The valve is operated by the starting or valve lever until the engine gets going. The eccentric rod is then hooked on.
American built engines often used poppet valves. The idea was the same but the valves needed a more complicated system of levers.

Good Sailing